D0151465

LETTERS ON ABSOLUTE PARALLELISM
1929-1932

ISBN 0-691-08229-4
L. C. 78-73832

PRINCETON UNIVERSITY PRESS
AND ACADÉMIE ROYALE DE BELGIQUE

Elie Cartan - Albert Einstein
Letters on Absolute Parallelism
1929-1932

Original text, English translation
by Jules LEROY and Jim RITTER

Edited by Robert DEBEVER

PRINCETON
UNIVERSITY PRESS
1979

6459-3551
PHYSICS

repl
XD80.4314

Copyright © 1979 by
The Estate of Albert Einstein
and The Estate of Elie Cartan

QC15
C27
1979
PHYS

Présentation

1979, année du centenaire de la naissance d'Albert Einstein. Au concert d'hommages que le monde scientifique international rend à cette occasion au savant illustre qui, créant une conception nouvelle de la gravitation, a consacré le meilleur de ses forces à mettre au point la théorie de la relativité générale, l'Académie Royale de Belgique est heureuse d'apporter une contribution originale grâce à la publication, par les soins de Robert Debever, d'un important inédit: le texte de la correspondance échangée de 1929 à 1932 entre Albert Einstein et Elie Cartan et dont le thème était le parallélisme absolu.

C'est en 1931 qu'Einstein a été élu membre associé de l'Académie Royale de Belgique mais ses rapports avec notre pays remontent à bien plus haut. Dès leur création, en 1911, et trois fois encore dans la suite, en 1913, en 1927 et en 1930, il présenta des communications précieuses et décisives aux réunions des Conseils de Physique Solvay; c'est là aussi qu'il eut l'occasion de poursuivre des discussions fructueuses, entamées par correspondance, avec des maîtres éminents comme nos compatriotes Théophile De Donder et Georges Lemaître qui firent passer les idées einsteiniennes dans leur enseignement. Faut-il rappeler enfin les contacts confiants et étroits que, notamment dans les années sombres qui virent la montée du nazisme, Einstein entretint avec la famille royale, particulièrement avec la Reine Élisabeth.

Elie Cartan, dont l'œuvre mathématique considérable a marqué profondément les dernières décennies, a été élu membre associé de notre Académie en 1937. Mais c'est depuis de longues années qu'il avait noué des relations scientifiques et amicales avec des collègues belges.

Autant de raisons qui ont incité notre Compagnie à s'associer avec enthousiasme au tribut de reconnaissance payé à la mémoire de l'illustre savant en unissant les voix de deux de ses associés les plus célèbres.

Maurice LEROY
Secrétaire perpétuel
de l'Académie Royale de
Belgique

v

Préface

La correspondance entre E. Cartan et A. Einstein au sujet du parallélisme absolu a son objet propre qui constitue la substance des lettres qui suivent et des notes qui les accompagnent. Elle est aussi limitée dans le temps à la période 1929-1932.

Il nous a paru opportun de tenter de mettre en lumière des préoccupations permanentes de deux personnalités à la fois très différentes mais très cohérentes chacune dans leur cheminement.

Nous le ferons à partir de la présente correspondance mais plus largement, en évoquant en partie l'œuvre d'Einstein et de Cartan et leurs prolongements.

On sera frappé dès l'abord du rythme de la correspondance. Dans sa partie la plus importante : en quelque douze semaines, de décembre 1929 à février 1930, vingt-six lettres seront échangées auxquelles il faut ajouter deux longues notes mathématiques de Cartan. On rêvera aussi à l'efficacité et à la rapidité du service postal !

Mais en même temps il faut évoquer brièvement la situation sociale de ces deux savants. 1929 est l'année du cinquantenaire d'Einstein. Depuis dix ans, Einstein est devenu une célébrité mondialement connue, le symbole du savant le plus éminent. L'année 1929 sera l'occasion de mille manifestations d'hommage, de remises de diplômes de doctorat honoris causa, d'adresses, d'interviews et aussi d'épisodes tragi-comiques dans les relations d'Einstein avec la municipalité de Berlin, comiques par les palinodies qui jalonnent cet « hommage » et tragiques par les relents d'antisémitisme et nazisme qui se précisent. Einstein outre son activité scientifique est aussi sollicité de toute part, par les mouvements et organisations en faveur de la paix, par les mouvements sionistes, pour citer ceux qui le concernent le plus.

L'annonce de la publication prochaine d'une nouvelle théorie de l'illustre savant avait causé sensation dans la grande presse, on y parlait de « découverte de l'énigme de l'univers ». Une équipe de physiciens était mobilisée à la rédaction du *New York Herald Tribune* pour aider à la composition d'un article à paraître basé sur la transmission par cablogramme du texte de la communication du 10 janvier

faite par Einstein devant l'Académie de Berlin et qui paraîtra le 30 janvier dans les *Sitzungsberichte der Preussischen Akademie*.

Le *Daily Chronicle* obtenait le 26 janvier une interview où Einstein déclarait : « Depuis des années c'est ma plus grande ambition que de résoudre en unité la dualité des lois naturelles. »

À partir de 1930 Einstein s'absentera plus souvent de Berlin pour visiter ou séjourner à Leyden, Cambridge, Londres, Oxford, Paris, Bruxelles, Pasadena...

Le départ définitif de Berlin sera de fin 1932, celui d'Europe de fin 1933 après un séjour de plusieurs mois en Belgique.

De son côté Elie Cartan assure un enseignement important en Sorbonne et à l'École de Physique et de Chimie, dirigée par son ami Paul Langevin.

Cartan conférencie en France et à l'étranger et poursuit avec une intensité prodigieuse son œuvre scientifique : il publie plus de cent pages par an en 1929, 1930, 1931 sans compter un ouvrage important de plus de trois cents pages sur la *Géométrie projective complexe*, texte écrit au départ de notes de son enseignement de Sorbonne du semestre 1929-1930 et repris et amplifié. Cartan est aussi à la tête d'une famille de quatre enfants. Le Professeur J. Dieudonné en parlait en ces termes le 18 mai 1939 à l'occasion du jubilé scientifique d'Elie Cartan : « ...*j'ai eu le privilège... de vous voir maintes fois au milieu des vôtres, dans cette admirable famille que vous avez fondée ; j'ai pu y sentir l'affection sans bornes qui en unit tous les membres, dans la vie comme dans la mort, et la communauté de pensée et de foi dans l'idéal le plus élevé, qui en fait un modèle aussi achevé, sur son plan propre, que votre œuvre scientifique l'est sur le sien* ». Cartan vit au Chesnay près de Versailles et passe chaque semaine de longues heures en chemin de fer.

On est confondu par l'activité et par l'ardeur des deux correspondants.

S'il s'attache à l'aspect extérieur de la correspondance, le lecteur sera sensible à l'expression très particulière de l'humour chez Einstein, à l'expression à la fois retenue, attentive et précise des idées et des sentiments chez Cartan.

S'il s'agit au début, d'une mise au point quant à la contribution de Cartan, très vite et surtout après le contact personnel de novembre 1929, les relations revêtiront à la fois l'intensité que nous avons évoquée et un caractère nettement plus personnel.

VIII

Bornons-nous à en évoquer quelques traits.

Einstein devant la difficulté d'exploiter comme il l'espérait sa théorie a cette image : « *Je suis comme un singe affamé qui après une longue recherche a trouvé une prodigieuse noix de coco et qui ne peut pas l'ouvrir et savoir ce qu'il y a dedans* » (13 février 1930). Cartan y fait écho : « *Est-il bien sûr que la noix de coco contienne quelque chose à l'intérieur ? On est devant un mur et les mathématiciens sont bien embarrassés pour percer une ouverture. On ne peut guère fonder d'espoir que sur un miracle de divination ; mais vous en avez déjà eu quelques-uns !* » (17 février 1930).

Le 16 mai 1932 Einstein écrit en s'excusant d'abuser de la bonté de son correspondant : « *Je me console avec l'illusion : peut-être prend-il aussi quelque plaisir à ce petit échange. Imaginez que nous soyons à nouveau jeunes tous deux et que je sois, votre élève certes zélé, mais peu doué.* » Cartan répond le 24 mai : « *Votre lettre me remplit à la fois de joie et de confusion. Sûrement j'éprouve du plaisir à notre petite correspondance ; s'il ne tenait qu'à moi je redeviendrais volontiers jeune sinon pour vous donner des leçons, du moins pour pouvoir mieux que je ne le puis maintenant suivre tout ce qui se fait de merveilleux en physique.* »

La recherche d'une théorie unitaire est très certainement une idée fondamentale, permanente d'Einstein. C'est la conviction profonde que l'unité des phénomènes naturels doit trouver son expression dans une théorie globalisante. Dans le premier mémoire de la série de travaux consacrés au parallélisme absolu il exprime « *l'espoir de parvenir à une construction logique qui unisse tous les concepts de champs physiques sous un même point de vue* ». Mais la réalisation ne trouvera pas sa fin du vivant d'Einstein. Les tentatives commencèrent dès 1918 avec H. Weyl et A. Eddington. En 1923 Einstein reprenait à peu de choses près le point de vue d'Eddington, en 1925 il croit avoir trouvé la vraie solution : « *Après une recherche inlassable durant les deux dernières années je crois avoir trouvé la vraie solution* ». Curieusement le travail de 1925 contient tous les ingrédients des essais futurs et pas seulement ceux d'Einstein. Tout d'abord une connexion affine générale, avec courbure et torsion, ensuite un champ tensoriel de densité deux fois contravariante, non symétrique. Une série de spécialisations permettent à Einstein de trouver à ce stade, en première approximation, des équations analogues aux équations de Maxwell. Le même espoir

renaîtra lors de la tentative de parallélisme absolu, encore que devenu prudent (le restera-t-il?), à Paris en novembre 1929 il dira: « *À l'heure actuelle cette nouvelle théorie n'est qu'un édifice mathématique, à peine relié par quelques liens très lâches à la réalité physique* ».

Au début de 1932 Einstein abandonnera sa tentative du parallélisme absolu pour la nouvelle théorie élaborée avec W. Mayer: « *J'ai l'espoir que cette théorie approche réellement la structure de l'espace physique* » (21 mars 1932) et l'on pourrait continuer sur ce ton jusqu'aux ultimes tentatives des champs deux fois covariants non symétriques des dernières années. Ce que nous voulons retenir de cela c'est une caractéristique très constante de la façon de travailler, de créer d'Einstein. S'il a beaucoup d'idées il n'en travaille qu'une seule à la fois, ne se laisse distraire par rien de sa méditation, est sourd aux travaux d'autrui, n'utilise que ce qui a été élaboré par lui-même. Si finalement il échoue, il repart dans une nouvelle direction avec la même foi, et le même enthousiasme.

La correspondance qui suit est très démonstrative à cet égard. Cartan a vite fait de voir et de montrer que plusieurs possibilités sont ouvertes, dans le cadre des espaces à parallélisme absolu. Un champ de torsion et une connexion affine (sans courbure) peuvent conduire à plusieurs systèmes d'équations ayant une signification « invariante » ou géométrique, Cartan les écrira dès décembre 1929. Entre-temps Einstein avait trouvé sa version à 22 équations, rien ne pourra l'en écarter et son intérêt dans la discussion avec Cartan est d'essayer d'appliquer les résultats mathématiques de Cartan non en géométrie différentielle proprement dite, mais dans la discussion de systèmes linéaires d'équations aux dérivées partielles du premier ordre. Le « degré de généralité » ([1]) découvert par Cartan, devrait permettre de justifier son propre choix parmi les solutions possibles.

La correspondance est très révélatrice à cet égard de la manière dont Einstein assimile le travail de quelqu'un d'autre, il lui faut le repenser à sa façon, en se trompant quelquefois, mais rien n'y fait et il revient à la charge, insiste, modifie sa présentation. Mais avant tout c'est son intuition, « son point de vue d'architecture », comme le note Paul Valéry après l'avoir entendu le 12 novembre 1929, qui le

---

1. Le degré de généralité est une notion précise à l'expression multiforme. Le lecteur se reportera à ce sujet à la note 5 de la lettre VII.

guide. Il avoue lui-même n'avoir rien compris de ce que Cartan lui avait expliqué du parallélisme absolu dès 1922 à Paris au cours d'une soirée chez Hadamard. En 1929 il sera sourd à l'allusion de Cartan relative à une relativité générale à torsion. De toutes les tentatives d'Einstein il faut retenir que l'espoir d'une théorie unitaire est toujours vivace. Au cours de ces dernières années malgré les nouveaux « champs physiques » pour reprendre l'expression d'Einstein, ceux des interactions fortes et faibles, malgré et peut-être à cause de la prolifération de particules élémentaires, des chercheurs de talent s'efforcent toujours d'élaborer un schéma globalisant: on l'appelle aujourd'hui « supergravité ». Deux problèmes fondamentaux concernant la gravitation restent à résoudre. La gravitation « pure » est, semble-t-il, parfaitement décrite par la théorie d'Einstein de 1915, mais comment la force de gravitation est-elle liée avec les autres types de forces et comment peut-elle être compatible avec les principes de la mécanique quantique? La théorie de la supergravité propose une nouvelle approche du problème de l'unification, l'avenir nous apprendra jusqu'où elle peut aller.

Nous sommes restés jusqu'ici sur l'orbite « Einstein ».

Qu'en est-il de « l'orbite Cartan »? D'abord la gloire du mathématicien est toujours confinée dans une aire plus confidentielle. Les « médailles Fields », équivalents symboliques du Nobel, n'existent que depuis 1936. Si beaucoup de mathématiciens célèbres sont précoces, très précoces et parfois précocement célèbres, d'autres tout aussi précoces élaborent une œuvre considérable qui ne révèle toute son ampleur que bien des années après. Le cas d'Elie Cartan en est un bon exemple, et peut-être un des derniers; car de nos jours la circulation des idées est beaucoup plus rapide, le nombre des chercheurs intéressés sans commune mesure avec ce qu'il était il y a 50 ans et plus.

Si la théorie des groupes de Lie est aujourd'hui un des piliers de la mathématique c'est bien à Sophus Lie qu'on le doit mais aussi et surtout à Cartan.

La théorie de Lie a un aspect algébrique qui a été l'objet de la thèse d'Elie Cartan. Cette thèse date de 1894, il est hautement significatif de savoir qu'elle a été rééditée *ne varietur* en 1933. Tous les grands problèmes de la classification des groupes simples et semi-simples y sont traités et résolus de manière quasi définitive. C'est encore Cartan qui, complétant sa thèse, a déterminé les groupes simples à paramètres

réels (1914) et en 1913 toutes les représentations linéaires irréductibles des groupes simples et semi-simples et découvert les spineurs. Un autre pilier de l'œuvre de Cartan est sa théorie des systèmes d'équations aux dérivées partielles où la préoccupation première était de fonder une théorie où n'interviennent que des notions et des opérations indépendantes de tout changement de variables, ce qui est la préoccupation première du géomètre. C'est ainsi que Cartan élabore, et doit élaborer sa théorie des systèmes d'équations aux différentielles totales ou systèmes de Pfaff. La richesse du contenu s'en révéla par les applications innombrables qu'il en a faites, que ce soit dans la théorie des groupes continus de Lie, dans celle des groupes infinis, dans de multiples questions de géométrie différentielle.

C'est la combinaison des résultats sur la représentation des groupes de Lie et l'intégration des systèmes de Pfaff qui conduisent Cartan à sa théorie des espaces généralisés et là nous retrouvons Einstein, non pas dans la présente correspondance, encore que ces deux chapitres en constituent la substance, mais en 1921-1922, à la création même du concept nouveau. Un passage de la notice de Cartan sur ses *Travaux scientifiques* rédigée en 1931 est très claire à ce sujet: « *Les espaces que j'ai imaginés... sont... une généralisation des espaces riemanniens, des espaces de Weyl et d'Eddington, mais à laquelle il était impossible de parvenir en suivant les idées directrices de Riemann, de M. Weyl et de M. Eddington. C'est ma conception de la structure des groupes continus qui m'a guidé et qui m'a permis de faire jouer à la notion de groupe un rôle fondamental dans un domaine où elle semblait exclue* ». Et il est important de souligner que c'est la rencontre de Cartan avec la théorie de la gravitation d'Einstein qui en sera l'occasion et les travaux que nous allons analyser paraîtront à quelques semaines à peine du séjour d'Einstein à Paris en fin mars 1922.

Cartan publie successivement un grand mémoire

1. Sur les équations de la gravitation d'Einstein. *Journal de Math. pures et appliquées*, t. 1, pp. 141-203...

suivi de deux notes aux *Comptes Rendus* de Paris

2. Sur une définition géométrique du tenseur d'énergie d'Einstein. *C.R. Paris*, t. 174, p. 437, 13 février.

3. Sur une généralisation de la notion de courbure de Riemann et les espaces à torsion. *Idem*. p. 593, 27 février.

Trois autres notes aux *Comptes Rendus* indiqueront les grandes voies nouvelles qui s'ouvrent après les découvertes de la note 3. L'intérêt de ce mémoire et des deux notes qui suivent est de nous faire assister à la naissance d'un concept nouveau avec la prescience de sa profonde richesse.

Le premier mémoire a d'ailleurs été complété sur épreuves après la présentation des notes aux *Comptes Rendus*. On y lit en effet « *Ce mémoire a été rédigé il y a plus d'un an. Depuis j'ai publié en février et mars derniers, dans les* COMPTES RENDUS DE L'ACADÉMIE DES SCIENCES (t. 174, pp. 437, 543, 734, 857, 1104) *des Notes relatives à une conception géométrique nouvelle des espaces non euclidiens. L'idée fondamentale de ces Notes est en germe, sous une forme mi-abstraite, mi-géométrique dans les premiers et derniers numéros de ce mémoire (1-6; 35-40).* »

Quel en était l'objet? Comme l'indique le titre, rechercher les conditions mathématiques qui rendent compte de la nécessité du tenseur d'Einstein

$$G_{ij} = R_{ij} - \frac{1}{2} R g_{ij},$$

combinaison du tenseur fondamental g_{ij}, du tenseur de Ricci R_{ij} et de la courbure scalaire R.

Quels sont les outils de Cartan: la théorie des représentations linéaires d'un groupe, et la théorie des équations aux différentielles totales pour traiter de l'équivalence des espaces riemanniens ou pseudo-riemanniens afin de dégager les notions réellement géométriques, ou indépendantes de la représentation analytique. Dans les paragraphes 1-6 auxquels Cartan fait allusion on trouve les formules

$$d\omega_i = \omega_k \wedge \omega_{ki}$$
$$\Omega_{ij} = d\omega_{ij} - \omega_{ik} \wedge \omega_{kj}$$

où ω_i sont les 1-formes de base d'un corepère, les ω_{ki} la 1-forme de connexion, Ω_{ij} la 2-forme de courbure. Les paragraphes 35-40 présentent les identités de Bianchi et une 3-forme à valeurs vectorielles à laquelle est précisément associé le tenseur G_{ij} et qui satisfait aux identités de conservation. Cette 3-forme est parfois connue sous le nom de courbure trivectorielle.

L'objet de sa note 2 sera précisément de donner une interprétation géométrique simple de cette 3-forme, comme « rotation », ici transformation de Lorentz, associée à un contour fermé tridimensionnel.

Le sens de la généralisation est exprimé dans la note 3, citons: « *Dans la note 2 j'ai fait intervenir la courbure par une certaine rotation s'appuyant sur le parallélisme de Levi-Civita. Il est possible d'en montrer la signification profonde en généralisant la notion même d'espace* ». À tout contour fermé sera associé non plus une rotation, mais un déplacement: rotation et translation, courbure et torsion, les formules ci-dessus prennent alors leur forme symétrique

$$\Omega_i = d\omega_i - \omega_k \wedge \omega_{ki}$$

$$\Omega_{ij} = d\omega_{ij} - \omega_{ik} \wedge \omega_{kj}$$

où Ω_i est la 2-forme de torsion. Si les premiers membres des équations ci-dessus sont nuls on retrouve les équations de structure du groupe des déplacements.

La possibilité d'aller beaucoup plus loin est évoquée d'un mot à la fin de la note 3, « les considérations précédentes... peuvent elles-mêmes se généraliser ». En effet le groupe des déplacements pourra être remplacé par d'autres: affins, projectifs, conformes...

Et le développement de la découverte suivra à un rythme effréné. C'est pendant encore 25 ans et dans plus de cent mémoires, notes, rapports de congrès, que Cartan poursuivra ses recherches. Enfin c'est après la seconde guerre mondiale, avec les progrès de la Topologie, de l'Algèbre que l'œuvre de Cartan sera à la fois systématisée, approfondie et développée.

Le premier grand mémoire consacré par Cartan à sa découverte est intitulé « Sur les variétés à connexion affine et la théorie de la relativité généralisée ». Partant du groupe de Galilée, Cartan présente la Mécanique classique comme espace-temps à connexion « galiléenne »; les lois fondamentales de la Mécanique apparaissent comme des conditions sur la 2-forme de courbure mise en relation avec la répartition spatio-temporelle de la matière et éventuellement comme des conditions sur la torsion s'il existait un moment cinétique élémentaire de la matière. Cette idée est de 1923, le spin ne sera mis en évidence expérimentalement qu'en 1925 et cet aspect du mémoire de Cartan restera en sommeil pendant près de quarante ans.

La Relativité générale s'introduit ensuite de manière naturelle si l'on substitue le groupe de Lorentz au groupe de Galilée. Signalons encore que dans ce mémoire le parallélisme absolu est présenté au passage comme propriété des espaces à connexion affine sans courbure.

Curieusement nous devons encore signaler un travail de Cartan en relation avec les travaux d'Einstein. Il s'agit d'une note posthume intitulée « La théorie unitaire d'Einstein et Mayer », note entièrement rédigée et qui fut révélée en 1954 par les éditeurs des *Œuvres complètes* d'E. Cartan.

Lorsqu'Einstein abandonnera sa tentative du parallélisme absolu il élaborera en effet avec W. Mayer une théorie sur une variété à quatre dimensions avec une fibration à 5 dimensions. La note de Cartan analyse le travail d'Einstein et Mayer et le caractérise comme géométrie induite d'une sous-variété totalement géodésique à quatre dimensions d'un espace à connexion euclidienne à cinq dimensions.

Nous ne savons pas ce qui explique pourquoi ce texte n'a pas été publié. Bien sûr les événements: prise du pouvoir par Hitler, installation d'Einstein à Princeton expliquent l'arrêt de la correspondance. Nous restons ici sur une interrogation.

Il y aura un échange de quelques lettres après la seconde guerre mondiale. Nous ne les avons pas jointes à la présente édition, elles sont sans lien avec les précédentes. Nous n'en retiendrons que quelques éléments.

D'abord pour évoquer la situation particulière d'Einstein au sein de la communauté des « physiciens théoriciens », la position qu'il a prise et défendue contre vents et marées vis-à-vis de la mécanique quantique et son interprétation probabiliste.

Le 7 janvier 1930 quand il écrit à Cartan pour souligner les difficultés inhérentes à sa recherche de solutions sans singularités des équations du champ il écrit: « *Le pire c'est que nos physiciens théoriciens ne veulent pas collaborer, au contraire ils m'injurient, parce qu'ils n'ont pas d'organe pour sentir le naturel du chemin suivi (exception faite pour Langevin!)* ».

Le 21 mars 1932 il ajoute: « *Je ne puis absolument pas me satisfaire du dogme de l'interprétation purement statistique des physiciens de la nouvelle génération, aussi séduisants que soient leurs arguments* ».

Enfin le 21 juin 1950 dans la dernière lettre d'Einstein à Cartan, après avoir fait allusion à sa théorie des g_{ij} non symétriques, il écrit: « *Il est cependant très difficile de prouver si toute cette théorie est conforme à la réalité physique. Il y a cependant une chose sur laquelle je n'ai aucun doute c'est que tout schéma physique basé fondamentalement sur le concept de probabilité se révélera en fin de compte inacceptable* ».

Des lettres d'après 1945 nous en retiendrons une cependant, écrite par Cartan à Einstein à l'occasion de son septantième anniversaire. En cette année du centième anniversaire d'Einstein cette lettre peut être conservée sans changement. Elle est datée du 1er mars 1949. En voici le texte :

« *Mon cher Collègue et cher Maître,*

C'est lundi prochain 14 mars, à l'occasion de votre soixante-dixième anniversaire, que les savants du monde entier, mathématiciens, physiciens et chimistes, fêteront votre jubilé. J'avais été prévenu que deux volumes d'articles et de mémoires seraient consacrés à la commémoration de ce jubilé.

Comme j'approche de mon quatre-vingtième anniversaire, je me suis senti incapable de faire paraître dans ces volumes un article dont la valeur scientifique soit digne de vous. C'est pour cela que je viens vous dire simplement mes sentiments d'affectueuse admiration pour vos travaux, qui ont bouleversé les notions les plus anciennes de l'humanité sur l'espace et le temps, et qui ont renouvelé complètement les principes de la Physique.

Permettez-moi de vous rappeler que j'ai eu l'honneur d'avoir avec vous une correspondance suivie après la publication de votre théorie unitaire du champ. Puisque j'en suis à rappeler mes souvenirs, je me vois encore avec vous dans le bureau de Paul Langevin à l'École de Physique et de Chimie. Je sais quelle admiration, pour ne pas dire quelle vénération, vous aviez pour lui, pour ses qualités scientifiques et humaines, pour la clarté bien rare avec laquelle il savait dominer les problèmes les plus difficiles. Je ne puis séparer sa mémoire de votre œuvre, qu'il avait contribué puissamment à faire connaître et à répandre.

C'est en pensant à Paul Langevin que je vous renouvelle mes sentiments d'admiration, avec l'espoir que vous vivrez encore assez longtemps pour assister au développement des progrès, peut-être encore imprévus, dont la Science trouvera la source dans votre œuvre elle-même.

Je vous prie d'agréer, cher Maître, l'expression de ma dévouée sympathie.

Elie CARTAN
95, boulevard Jourdan
Paris (14e)

* * *

Mes remerciements les plus vifs s'adressent au professeur Henri Cartan. C'est lui qui nous a communiqué les textes rendus publics aujourd'hui. Il s'est montré immédiatement favorable à une publication à l'initiative de notre Académie. Il nous a aidé tout au long de l'entreprise et nous lui devons la très belle photo de son père reprise en tête de la liseuse.

Je remercie aussi vivement la succession Einstein et son administrateur Monsieur Otto Nathan qui nous a accordé les autorisations nécessaires et mis en rapport avec la *Princeton University Press*. La coédition s'est réalisée dans un parfait esprit de coopération et d'entraide aux différentes étapes de la réalisation; nous en remercions très chaleureusement le directeur Monsieur Herbert S. Bailey.

Je remercie aussi particulièrement le Professeur Francis Perrin qui a bien voulu me communiquer des documents d'époque en rapport avec la présente correspondance.

J'ai bénéficié de l'excellente collaboration des traducteurs Monsieur Jim Ritter du King's College de Londres et de Monsieur Jules Leroy de l'Université Libre de Bruxelles. Ils ont exécuté un travail difficile avec beaucoup de gentillesse et de soin et contribué à la mise au point finale du volume; je leur en suis très reconnaissant.

Je remercie la famille Koch-Ferrard, dépositaire d'importants documents, qui nous a fourni la photo d'Einstein de la liseuse ainsi que Madame H. Diserens qui a réalisé le projet nécessaire à la réalisation de la page de dos de la liseuse.

Merci à Monsieur Jean-Luc De Paepe, attaché scientifique de l'Académie, qui n'a jamais ménagé ses efforts pour nous aider en différentes occasions.

Enfin je remercie la maison Duculot pour avoir fait diligence et apporté tous ses soins à l'impression du présent volume.

R. Debever

Note liminaire au sujet des traductions

La traduction en anglais des en-têtes et fins de lettre du texte français présentait des difficultés souvent insurmontables. Nous avons pour les lettres de Cartan reproduit l'expression française quand cela s'imposait et renvoyé le lecteur au texte original quand aucun équivalent convenable n'a été trouvé. La traduction anglaise de l'allemand s'est révélée plus facile, plus brève, sans qu'il n'y ait cependant avantage pour le lecteur intéressé à se reporter au texte original.

Preliminary note on the translation

The translation into English of the French salutations and closings of the letters in the volume often presented insurmountable difficulties. When this has occurred in Cartan's letters we have simply reproduced the original French and we refer the reader to the original text. The translation from the German into English has proven to be considerably easier and briefer; even so the interested reader can refer with profit to the original text.

R. Debever

1

Table des lettres

TABLE DES LETTRES

I

Le Chesnay (S. et O.)
27 avenue de Montespan [1],
le 8 mai 1929

Monsieur et illustre Maître,

Je m'excuse de prendre quelques instants de votre temps si précieux pour la science; c'est sur le conseil de mon ami Langevin que je me décide à vous écrire.

Dans vos notes récentes des *Sitzungsberichte* consacrées à une nouvelle théorie de la relativité généralisée, vous avez introduit dans un espace riemannien la notion de *Fernparallelismus* [2] [16-18]. Or la notion d'espace riemannien doué d'un *Fernparallelismus* est un cas particulier d'une notion plus générale, celle d'espace à connexion euclidienne, que j'ai indiquée succinctement en 1922 dans une note des *Comptes rendus* (t. 174, pp. 593-595) [3] parue au moment où vous faisiez vos conférences au Collège de France [3]; je me rappelle même avoir, chez M. Hadamard, essayé de vous donner l'exemple le plus simple d'un espace de Riemann avec *Fernparallelismus* en prenant une sphère et en regardant comme parallèles deux vecteurs faisant le même angle avec les méridiennes qui passent par leurs deux origines: les géodésiques correspondantes sont les loxodromies. Cet exemple est cité aussi dans un article: Sur les récentes généralisations de la notion d'espace (*Bull. Sciences math.* 48, 1924, pp. 294-320) [5].

---

1. Le Chesnay est une localité de la banlieue Nord de Versailles à vingt kilomètres du centre de Paris. Elie Cartan y a vécu de 1917 à 1936.

2. Les renvois entre crochets se rapportent à la bibliographie des travaux directement liés au texte de la correspondance que l'on trouve en fin du volume. Einstein emploie le terme *Fernparallelismus* dont l'équivalent anglais est le terme, toujours utilisé, de *distant parallelism*, en français parallélisme à distance. Cartan emploie systématiquement *parallélisme absolu*. On trouve aussi, utilisé en anglais: *absolute parallelism*, chez différents auteurs. C'est pourquoi nous n'avons pas uniformisé les traductions anglaises.

4

A I

Le Chesnay (S. et O.)
27 avenue de Montespan [1],
8 May 1929

Monsieur et illustre Maître,

I apologize for taking a few moments of your time which is so precious for science, I decided to write to you on my friend Langevin. advice.

In your recent articles in the *Sitzungsberichte* devoted to a new theory of generalized relativity, you introduced, the notion of " *Fernparallelismus* " [2] [16-18] in a Riemannian space. Now, the notion of Riemannian space endowed with a *Fernparallelismus* is a special case of a more general notion, that of space with a Euclidean connection, which I outlined briefly in 1922 in an article in the *Comptes Rendus* (vol. 174, pp. 593-595) [3], published when you gave your lectures at the *Collège de France* [3]; I even remember trying, at Mr Hadamard's home, to give you the simplest example of a Riemannian space with *Fernparallelismus* by regarding two vectors within a sphere making the same angle with the meridian lines passing through their origins as parallel: the corresponding geodesics are the rhumb lines. This example is quoted in an article: " Sur les récentes généralisations de la notion d'espace " (*Bull. Sciences math.* 48, 1924, pp. 294-320) [5].

---

3. Il s'agit de quatre conférences données par Einstein au Collège de France à l'invitation de P. Langevin. La première conférence eut lieu le vendredi 31 mars 1922. Ce fut un événement à la fois scientifique et politique.

On lira par exemple à ce sujet, en français Banesh Hoffman: *Albert Einstein, créateur et rebelle*, éd. du Seuil, Paris 1972, (trad. de l'anglais) p. 163 avec une photo dans le texte relative à sa conférence au Collège de France. Le lecteur pourra aussi se reporter à la grande biographie de R. W. Clark: *Einstein, The life and Times*, Avon Printing, New York, 1972, pp. 354-355.

Avec la terminologie que j'ai introduite les espaces à connexion euclidienne comportent une *courbure* et une *torsion*; dans les espaces où le parallélisme est défini à la Levi-Civita, la torsion est nulle; dans les espaces où le parallélisme est absolu (*Fernparallelismus*) la courbure est nulle, ce sont donc des espaces sans courbure et à torsion. J'ai fait dans un long mémoire paru dans les *Annales de l'École normale* [4] (surtout t. 42, 1925) et intitulé: « Sur les variétés à connexion affine et la théorie de la relativité généralisée » une étude systématique des tenseurs auxquels donne lieu soit la courbure soit la torsion: l'un de ceux que fournit la torsion a précisément tous les caractères mathématiques du potentiel électromagnétique.

Les variétés riemanniennes avec *Fernparallelismus* jouent un rôle important dans la théorie des groupes. J'ai étudié tout cela dans un autre long mémoire: La Géométrie des groupes de transformations (*J. de math. pures et appliquées*, t. 6, 1927, pp. 1-119) [7]; dans l'espace représentatif des transformations d'un groupe continu, il existe en effet deux connexions affines sans courbure (*Fernparallelismus*) remarquables. La condition pour que le tenseur introduit par le *Fernparallelismus* soit à coefficients constants est précisément que l'espace considéré soit l'espace représentatif d'un groupe (simple ou semi-simple si l'espace est riemannien).

Je ne veux pas vous ennuyer plus longtemps par l'énumération de travaux où la notion de torsion est utilisée. Je me permets simplement de vous envoyer le texte de deux conférences faites l'une à Toronto en 1924 [6], l'autre à Berne en 1927 [8], où j'ai exposé sans aucune formule ma théorie générale, qui dépasse du reste de beaucoup le domaine de la géométrie riemannienne; je me permettrai de vous signaler particulièrement dans la première conférence ce qui est dit pages 92-93 sur l'espace géométrique de votre première théorie de la relativité; et aussi dans la seconde p. 209, où je parle explicitement du parallélisme absolu et p. 217 où j'introduis les deux parallélismes absolus d'un espace de groupe [4].

---

4. Les remarques sur lesquelles E. Cartan attire l'attention se trouvent aux pages 902 et 903, *O.C.*, III, 1 pour le texte de la conférence de Toronto: elles portent sur une caractérisation soit projective, soit conforme des lois de la gravitation d'Einstein, en terme de groupe d'holonomie.

Pour le texte de conférence à Berne, *O.C.*, I, 2, p. 850. E. Cartan définit le parallélisme absolu comme connexion affine sans courbure, et à la page 858 les deux parallélismes

8 MAY 1929

In my terminology, spaces with a Euclidean connection allow of a *curvature* and a *torsion*: in the spaces where parallelism is defined in the Levi-Civita way, the torsion is zero; in the spaces where parallelism is absolute (*Fernparallelismus*) the curvature is zero, thus these are spaces without curvature and with torsion. In a long paper published in the " *Annales de l'École normale* " (especially vol. 42, 1925) [4] entitled " Sur les variétés à connexion affine et la théorie de la relativité généralisée "; I have systematically studied the tensors which arise from either the curvature or the torsion: one of those given by the torsion has precisely all the mathematical characteristics of the electromagnetic potential.

Riemannian manifolds with *Fernparallelismus* play an important role in group theory. I examined this in another long paper: " La Géométrie des groupes de transformations " (*J. de math. pures et appliquées* vol. 6, 1927, pp. 1-119) [7]; In the space representing the transformations of a continuous group, there exist two remarkable affine connexions without curvature (*Fernparallelismus*). For the tensor introduced by the *Fernparallelismus* to have constant coefficients the space considered should be the representative space of a group (a simple or semi-simple group if the space is Riemannian).

I shall not bother you any longer with the enumeration of papers where the concept of torsion is used. Allow me simply to send you the text of two lectures I gave one in Toronto in 1924 [6], the other in Bern in 1927 [8], in which I outlined, without formulae, my general theory (which, in any case, goes far beyond the domain of Riemannian geometry). Allow me to draw especially your attention to my first lecture and in particular to what is said on page 92-93 concerning the geometrical space of your first relativity theory; and also to the second one, p. 209, where I explicitly discuss absolute parallelism and to p. 217 where I introduce the two absolute parallelisms of a group space [4].

---

absolus qui s'introduisent très naturellement dans les espaces de groupe de Lie: si g_1, g_2 et g'_1, g'_2 sont deux paires de points de l'espace du groupe les conditions

$$g_2 g_1^{-1} = g'_2 g'_1{}^{-1} \quad \text{et} \quad g_1^{-1} g_2 = g'_1{}^{-1} g'_2$$

définissent deux équipollences et par suite deux parallélismes absolus sur l'espace du groupe.

Veuillez excuser, Monsieur et illustre Maître, cette lettre trop longue et agréer l'expression de mes sentiments de haute considération.

> E. Cartan,
>
> 27 avenue de Montespan
> Le Chesnay (Seine et Oise)

Le mémoire sur les variétés à connexion affine et la théorie de la relativité généralisée a paru dans les *Annales de l'École normale* t. 40, 1923, pp. 325-412; t. 41, 1924, pp. 1-25; t. 42. 1925, pp. 17-88.

8 MAY 1929

Please forgive me *Monsieur et illustre Maître* for too long a
letter...

E. Cartan,

27 avenue de Montespan
Le Chesnay (Seine et Oise)

The paper on manifolds with affine connection and the theory of
generalized relativity was published in the "*Annales de l'École
normale*" vol. 40, 1923, pp. 325-412; vol. 41, 1924, pp. 1-25; vol. 42,
1925, pp. 17-88.

II

Berlin 10.V.29

Lieber Herr Kollege!

Ich sehe in der That ein, dass die von mir benutzten Mannigfaltigkeiten in den von Ihnen studierten als Spezialfall enthalten sind. Auch die Herren Eisenhart [27] (Princeton) und Weitzenböck [29] (Saaren) haben die mathematische Grundlage meiner neuen Theorie bereits teilweise vor mir dargelegt [1]. Letzterer hat in einer zu unserer Akademie *Sitz. Ber.* 1928, XXVI, gedruckten Abhandlung ein — wie es schien vollständiges — Literatur -Verzeichnis der einschlägigen mathematischen Arbeiten angegeben; *dabei hat er aber auch Ihre Arbeiten übersehen.* Dies muss nun wieder gut gemacht werden. Ich bin aber ein bischen ratlos, wie ich es machen soll, um alle gerechten Anspruche zu befriedigen.

Ich habe gestern der *Zeitschrift für Physik* eine zusammenfassende Arbeit über den Gegenstand eingereicht, in der ich den Gegenstand ausführlich behandelt habe, ohne mich auf irgendwelche früheren Publikationen (auch nicht auf meine eigenen) zu stützen [2]. Dieser Arbeit könnte ich eine Nachschrift beifügen, in welcher die mathematische Vorgeschichte der Theorie behandelt wird. Ich habe aber Angst, dass ich bei Ausführung dieses Vorhabens wieder nicht allen Beteiligten werde ein befriedigender Weise gerecht werden können. Deshalb mache ich Ihnen folgenden Vorschlag: Schreiben Sie über diese mathematische Vorgeschichte eine kurze Charakteristik, die wir meiner neuen zusammenfassenden Arbeit anheften, natürlich unter Ihrem Namen, aber mit meiner Arbeit zu einem Ganzen vereinigt. Sie würden mir einen grossen Gefallen erweisen, und wir würden damit ein gutes Beispiel geben, wie derartige Prioritätsfragen in einer würdigen und sympathischen Weise behandelt werden können.

Ich habe Ihre damaligen Ausführungen in Paris gar nicht verstanden; noch weniger war mir klar, wie sie für die physikalische

---

1. La bibliographie de R. Weitzenböck comporte quatorze références, le nom d'E. Cartan n'y figure pas.

A II

Berlin 10.V.29

Dear Colleague,

I see, indeed, that the manifolds used by me are a special case of those studied by you. Eisenhart [27] (at Princeton) and Weitzenböck [29] (at Saar) also partially laid out the mathematical foundations oı my new theory before I did [1]. The latter, in an article published in our Academy's *Sitz. Ber.* 1928, XXVI, has given a (supposedly complete) bibliography of relevant mathematical works; *but he has quite overlooked your work.* This must now be rectified. But I am a bit puzzled as to how I should do it so as to satisfy all just claims.

Yesterday I sent off a review article on the subject to the *Zeitschrift für Physik*, in which I dealt with the topic in detail but without relying on any earlier publications (not even my own)[2]. I could add a postscript to this article which would discuss the mathematical antecedents of the theory. But I am afraid that even the execution of this plan would still not be a satisfactory way of doing justice to all the parties concerned. Therefore, I make the following suggestion to you: write a short analysis of the mathematical background which we will append to my new review, under your name of course, but integrally joined to my article. You should be doing me a great favour and, at the same time, we should be giving a good example of how similar questions of priority might be handled in a dignified and sympathetic way.

I didn't at all understand the explanations you gave me in Paris; still less was it clear to me how they might be made useful for physical theory. I first remarked last year that it would be quite natural to add the hypothesis of distant parallelism to the Riemann metric. But only in the last few months have I realized that this actually leads to a theory which corresponds to the hitherto existing knowledge of

---

2. Einstein abandonnera ce projet et donnera en août 1929 l'article qui paraîtra avec celui de Cartan aux *Mathematische Annalen* [20,9].

Theorie nutzbar gemacht werden könnten. Erst voriges Jahr bemerkte ich, dass es ganz natürlich sei, der Riemann-Metrik die Hypothese des Fernparallelismus zu adjungieren. Dass dies aber wirklich zu einer dem bisherigen physikalischen Wissen über die physikalischen Qualitäten des Raumes entsprechenden Theorie, d.h. zu brauchbaren und durch die formalen Grundlagen beinahe eindeutig bestimmten Feldgleichungen führe, das erkannte ich erst in den letzten paar Monaten.

Ich sende Ihnen meine bisher in der Akademie über den Gegenstand veröffentlichten Arbeiten. Die zweite über die angenäherten Feldgleichungen leidet noch an dem Übelstand, dass bei der dort getroffenen Wahl für die Hamilton-Funktion ein zentralsymmetrisches elektrisches Feld unmöglich wäre [17]. Die dritte ist leider vergriffen, sodass ich Ihnen mein einziges Exemplar senden muss (zur einheitlichen Feldtheorie) [18]. Ich bitte Sie, mir dieses Exemplar, sowie die Arbeit von Weitzenböck, von der ich auch nur das eine Exemplar habe, wieder zurückzusenden. Die richtige Lösung des Problems findet sich erst in der letzten Arbeit.

Indem ich Sie bitte, mein versehentliches Plagiat zu entschuldigen und mir zu helfen, die Angelegenheit in schöner Weise in Ordnung zu bringen unter Berücksichtigung aller Verdienste bin ich mit freundlichen Grüssen

Ihr

A. Einstein

P.S. Ich wurde Ihnen die Korrektur der neuen Arbeit senden, sobald ich sie erhalte [19].

the physical properties of space, i.e. to a useful set of field equations, almost uniquely determined by formal considerations.

I am sending to you my articles on the subject, published so far by the Academy. The second, on the approximate field equations, suffers, however, from the drawback that, with the choice made there for the Hamiltonian, a spherically symmetric electric field is impossible [17]. The third, unfortunately, is out of print so I am forced to send you my only copy (on the unified field theory) [18]. I beg you to send back this copy, as well as the article by Weitzenböck, of which I also have only the one copy. The correct solution of the problem first appears in the last article.

Asking you to forgive my inadvertent plagiarism and to help me settle the matter satisfactorily to everyone's benefit, with best wishes I am

Yours,

A. Einstein

I will send you the proofs of the new article as soon as I receive them 19].

III

Le Chesnay (Seine et Oise),
le 15 mai 1929

Monsieur et illustre Maître,

Veuillez m'excuser de ne pas avoir répondu plus tôt à votre lettre du 10 mai: je ne l'ai trouvée qu'aujourd'hui à la Sorbonne. Je vous retourne les deux tirages à part demandés, celui de votre dernière note et celui de Weitzenböck. Le silence de M. Weitzenböck à mon égard est un peu curieux car il indique dans sa bibliographie une note de Bortolotti dans laquelle il se réfère plusieurs fois à mes travaux.

J'accepte très volontiers votre proposition de faire, comme complément à votre prochain mémoire de la *Zeitschrift für Physik*, un court historique du parallélisme absolu. Bien entendu il ne s'agirait que de géométrie, en laissant plutôt à l'arrière-plan tout l'appareil analytique. Il me semble qu'il y aurait intérêt à indiquer les principaux problèmes qu'on s'est posés au sujet du parallélisme absolu, en montrant les relations qu'ils présentent avec certains problèmes de mécanique ou avec la théorie des groupes continus. Verriez-vous un inconvénient à ce que j'indique brièvement la théorie géométrique plus générale dans laquelle peut s'intégrer la notion de parallélisme absolu comme cas particulier? Je ne voudrais pas vous envoyer quelque chose qui dépasse les proportions que vous désirez.

Je vous remercie de vouloir bien me communiquer les épreuves de votre mémoire à mesure que vous les aurez; c'est avec grand plaisir que je les lirai.

Veuillez agréer, Monsieur et illustre Maître, l'expression de mes sentiments de dévouement et d'admiration.

E. Cartan

A III

Le Chesnay (Seine et Oise)
15 May 1929

Monsieur et illustre Maître,

Please excuse me for not having answered your letter of May 10 sooner: I only found it today, at the Sorbonne. I am returning the two requested reprints, your last article and that of Weitzenböck. M. Weitzenböck's silence is a bit strange for he lists in his bibliography a note of Bortolotti which refers several times to my work.

I accept with great pleasure your proposal to write, as a companion to your next article in the *Zeitschrift für Physik*, a short historical review on absolute parallelism. It is understood that it will only concern geometry, and relegate rather the whole analytical apparatus to the background. It seems to me that it would be interesting to indicate the main problems with absolute parallelism, showing their relationship to certain poblems in mechanics or to the theory of continuous group. Would you object if I briefly indicate the more general geometrical theory into which the notion of absolute parallelism can be integrated as a special case? I do not want to send you anything out of proportion.

I thank you for your willingness to send me the proofs of your article when you receive them; I shall read them with great pleasure...

E. Cartan

IV

Le Chesnay (Seine et Oise)
27 avenue Montespan,
le 24 mai 1929

Monsieur et illustre Maître,

Je vous envoie ci-inclus la notice historique sur le parallélisme absolu [9]. Je vous serais reconnaissant de vouloir bien y jeter un coup d'œil et me dire les modifications soit de fond soit de forme que vous jugeriez désirables. S'il y a des changements de rédaction à apporter, il est inutile de me renvoyer le manuscrit; j'en ai le double.

Veuillez agréer, Monsieur et illustre Maître, l'expression de mes sentiments les plus cordialement dévoués.

E. Cartan

Je vous serais reconnaissant de m'écrire de préférence, le cas échéant, au Chesnay.

A IV

Le Chesnay (Seine et Oise)
27 avenue de Montespan,
24 May 1929

Monsieur et illustre Maître,

I enclose the historical review of absolute parallelism [9]. I would be grateful if you would look at it and tell me the changes in either substance or in form that you think desirable. If changes have to be made, there is no point in sending me back the manuscript, I have a duplicate...

E. Cartan

I would be grateful to you for writing me, if need be, at Le Chesnay.

V

Albert Einstein

Berlin W. den 25 August 1929,
Haberlandstr. 5

Herrn Prof. E. Cartan
27, avenue de Montespan
Le Chesnay (S. & O.)

Lieber Herr Cartan,

Entschuldigen Sie bitte mein langes Schweigen. Dieses wurde verursacht durch viele Zweifel an der Richtigkeit des einge-schlagenen Weges. Nun aber bin ich soweit gekommen, dass ich die einfachste gesetzliche Charakterisierung einer Riemann-Metrik mit Fernparallelismus, welche für die Physik in Betracht kommen kann, gefunden zu haben überzeugt bin [1]. Ich schreibe nun die Arbeit für die *Mathematischen Annalen* und werde mir erlauben, die Ihrige der mei-nigen anzufügen, wie wir es verabredet haben. Die Publikation soll deswegen in den *Mathematischen Annalen* geschehen, weil einstweilen nur die mathematischen Zusammenhänge untersucht werden, nicht aber deren Anwendung auf die Physik.

Es grüsst Sie herzlich
Ihr

A. Einstein

---

1. C'est en effet dans le mémoire aux *Math. Annalen* [20] qu'Einstein donnera les équations de champ auxquelles il se tiendra désormais dans le contexte des espaces à parallélisme absolu.

Il s'agit de 22 équations construites à partir du tenseur de torsion $\Lambda^{\alpha}_{\beta\lambda}$, six équations notées

$$F_{\alpha\beta} \equiv \Lambda^{\mu}_{\alpha\beta;\mu} = 0$$

25 AUGUST 1929

A V

Albert Einstein

Berlin W. 25 August 1929,
Haberlandstr. 5

Professor E. Cartan
27, av. de Montespan
Le Chesnay (S. & O.)

Dear M. Cartan,

Please forgive my long silence. This has been caused by many doubts as to the correctness of the course I have adopted. But now I have come to the point that I am persuaded I have found the simplest legitimate characterization of a Riemann metric with distant parallelism that can occur in physics [1]. I am now writing up the work for the *Mathematische Annalen* and would like to be permitted to add yours to mine as we have agreed. The publication should appear in the *Mathematische Annalen* because, for the present, only the mathematical implications are explored and not their application to physics.

Kind regards,

Yours,

A. Einstein.

---

et 16 équations notées

$$G^{\alpha\beta} \equiv \Lambda^{\beta}_{\underline{\alpha}\,\mu;\mu} + \Lambda^{\rho}_{\underline{\alpha}\,\mu}\Lambda^{\beta}_{\mu\rho} = 0$$

où ; désigne la dérivation covariante et où les indices soulignés sont à mettre en position contravariante.

Nous donnons ici les équations sous leur forme originale car elles seront à la base des discussions qui occuperont les lettres ultérieures.

VI

Albert Einstein

Berlin W.
den 13 September 1929,
Haberlandstr. 5

Herrn Prof. Dr. E. Cartan
Le Chesnay (S. & O.)
27, avenue de Montespan

Lieber Kollege!

Heute gehen unsere beiden Arbeiten an die *Mathematischen Annalen* ab. Es hat solange gedauert, weil der Redakteur, Professor Blumenthal, auf einer Ferienreise war und infolgedessen nicht geantwortet hat. Er war besonders erfreut darüber, eine Originalabhandlung von Ihnen zu bekommen. Leider schreibt er mir, dass es bis zum Erscheinen der Abhandlung ein halbes Jahr dauern wird [1].
In der Hoffnung, Sie Anfang November in Paris zu sehen [2], bin ich

mit freundlichen Grüssen
Ihr

A. Einstein

---

1. Des lettres échangées entre Einstein et O. Blumenthal précisent les circonstances de cette publication.
C'est le 19 août qu'Einstein propose l'envoi de deux textes, l'un d'une quinzaine de pages de lui-même sur sa nouvelle théorie, et l'autre d'E. Cartan dont il demande l'impression à la suite du sien. Il s'inquiète de savoir si un texte en français peut paraître aux *Annalen* et aussi du temps nécessaire à la parution.
Le Prof. Blumenthal répond le 9 septembre, se réjouit de la proposition d'Einstein et accepte le manuscrit de Cartan, ajoutant que Cartan serait le bienvenu comme collaborateur par lui-même. Pour le délai il doit demander six mois de patience. Le 13 septembre Einstein envoie les deux manuscrits non sans faire remarquer que le délai lui paraît dommageable; « La physique a un autre rythme que les mathématiques ». Finalement

13 SEPTEMBER 1929

A VI

Albert Einstein

Berlin W.
13 September 1929,
Haberlandstr. 5

Prof. E. Cartan
Le Chesnay (S. & O.)
27, avenue de Montespan

Dear Colleague,

Today both our articles go off to the *Mathematische Annalen*. It has taken so long because the Editor, Professor Blumenthal, was on holiday and consequently had not replied. He was especially delighted to receive an original work of yours. Unfortunately, he writes to me, the articles will not appear for another six months [1].

In the hope of seeing you in Paris at the beginning of November [2], I am, with best wishes,

Yours,

A. Einstein

les textes paraîtront dans le numéro du 20 février 1930 des *Math. Annalen*, avec la mention « accepté le 19 août 1929 ».

2. C'est le samedi 9 novembre 1929 que se tiendra la cérémonie annuelle de réouverture de l'Université de Paris à l'occasion de laquelle eut lieu une proclamation solennelle de nouveaux docteurs *honoris causa*. Dans le grand amphithéâtre de la Sorbonne, sous la présidence du Recteur Charléty, devant une assistance de deux mille spectateurs le professeur Charles Maurain a énuméré les titres éclatants des nouveaux docteurs: A. Einstein, Sir Jencks (Fac. de Droit de Londres), Fr. Cumont (Historien et archéologue, Université de Gand), Dr. César Roux (Chimiste, Univ. de Lausanne) et du Prof. Mosiecki, Président de la République polonaise. On peut voir une photographie de cette séance dans *Albert Einstein* par Hilaire Cuny, éd. Seghers, Paris 1961, hors texte p. 96.

VII

Le Chesnay (S. et O.)
27 avenue de Montespan,
le 3 décembre 1929

Cher et illustre Maître,

Voici bientôt plus de trois semaines que vous avez quitté Paris et je ne vous ai pas encore donné signe de vie. Ce n'est pas que je n'aie beaucoup réfléchi à la suite de la conversation que nous avions eue chez Langevin [1]. Je vous avais posé des questions de nature mathématique ou physique qui ont pu vous sembler un peu naïves, mais c'est que je tenais à partir d'une base sûre et à pouvoir poser d'une manière précise le problème mathématique que vous vous étiez proposé. Je me permets de vous envoyer ci-inclus un exposé sommaire de la manière dont je conçois la question; je m'excuse d'avance d'entrer dans certains détails qui vous paraîtront peut-être un peu oiseux, mais c'est afin de vous faire comprendre mon point de vue; si je faisais fausse route, je désirerais que vous me le disiez. Je vous donne quelques détails sur la théorie des systèmes en involution que j'ai fondée il y a quelque trente ans et qui me semble tout à fait appropriée au problème que vous avez posé [2].

---

1. Le voyage à Paris fut l'occasion pour Einstein de faire deux conférences à l'Institut Henri Poincaré.

La première le vendredi 8 novembre à 17 h 30 comme en témoigne un billet d'Einstein à son ami Solovine (cf. A. Einstein. *Lettres à Maurice Solovine*. Gauthier-Villars, Paris, 1956, pp. 50-51).

La seconde le 12, suivie d'une discussion, est évoquée par P. Valéry; « 12 9bre, à 5 h 30. Conf[érence] d'Einstein. Je suis très intéressé vers la fin. Il se montre un grand artiste. » En renvoi, Paul Valery a noté: « Einstein: La distance entre la réalité et la théorie est telle qu'il faut trouver des points de vue d'architecture. Conférence, discussion du 12 ». [Paul Valéry, *Cahiers*, II, Bibliothèque de la Pléiade, Gallimard, Paris 1974, p. 875].

Le texte des conférences en français, rédigé par Al. Proca a été publié dans le premier fascicule des *Annales de l'Institut Henri Poincaré* [21].

Le séjour à Paris fut l'occasion d'un contact personnel entre Einstein et Cartan. À l'occasion du *Jubilé Scientifique d'Elie Cartan*, le 18 mai 1939 Langevin prononçait une allocution où il disait notamment: « Je veux seulement évoquer un souvenir personnel, celui des conversations qu'Einstein, lors de sa dernière visite à Paris, il y a bientôt dix

FACULTÉ DES SCIENCES   UNIVERSITÉ DE PARIS

GÉOMÉTRIE SUPÉRIEURE

Le Chesnay (S. et O.) 29 av. de Montespan

Paris, le 3 décembre 1929

Cher et illustre Maître,

Voici bientôt plus de trois semaines
que vous avez quitté Paris et je ne vous ai
pas encore donné signe de vie. Ce n'est pas
que je n'aie beaucoup réfléchi à la suite de
la conversation que nous avons eue chez
Langevin. Je vous avais posé des questions de
nature mathématique ou physique qui
ont pu vous sembler un peu naïves, mais
c'est que je tenais à partir d'une base sûre
et à prévoir sous d'une manière précise le
problème mathématique que vous
vous étiez proposé. Je me permets de
vous envoyer ci-inclus un exposé sommaire
de la manière dont je conçois la question;
je m'excuse d'avance d'entrer dans
certains détails qui vous paraîtront peut-
être un peu oiseux, mais c'est afin de
vous faire comprendre mon point de vue;

A VII

Le Chesnay (S. et O.)
27 avenue de Montespan,
3 Decembre 1929

Cher et illustre Maître,

It will soon be more than three weeks since you left Paris and still I have shown no sign of life. It's not that I haven't been thinking deeply about the conversation we had at Langevin's home [1]. If I asked you questions of a mathematical or physical nature that may have appeared a bit naive, it is because I wanted to start from a secure base and to be able to formulate your mathematical problem in a precise way. I enclose a brief account of the way in which I see the question; I apologize in advance for certain details that may seem a bit irrelevant, but this is because I want to help you understand my point of view; if I have taken the wrong road, I'd like you to tell me. I shall give you a few details of the theory of systems in involution which I discovered some thirty years ago that seems quite appropriate to your problem [2].

I have also made a few calculations without being able to decide if your solution is the best of all others. Another possible solution, one which contains two cosmological constants, consists in

---

ans, me demanda de lui ménager avec vous et dont j'ai conservé un émouvant souvenir. En pensant à vous, ces jours-ci, et en ouvrant l'un de vos Mémoires, j'ai retrouvé des feuilles où Einstein et vous avez, au cours de la discussion, mêlé vos écritures en confrontant, pour vous mieux entendre, les langages quelque peu différents que vous utilisiez » (Gauthier-Villars, Paris, 1939, pp. 31-32).

2. C'est en effet en 1899 que Cartan publie son premier mémoire sur ce sujet [1], suivi de [2].

Ces résultats seront à la base de la théorie des groupes infinis d'E. Cartan et trouveront bien des prolongements en géométrie différentielle.

La théorie de Cartan généralisée aux idéaux différentiels de degré supérieur à 2 par Kähler est aujourd'hui la matière du théorème de Cartan-Kähler. On peut en lire une présentation et démonstration dans J. Dieudonné: « *Éléments d'analyse*, t. 4, Gauthier-Villars, Paris 1971, ou *Treatise on Analysis*, vol. IV, Academic Press, New York and London 1974.

On lira aussi E. Cartan: *Les systèmes différentiels extérieurs et leurs applications géométriques*, Hermann, Paris, 1945.

J'ai aussi fait quelques calculs, mais sans arriver à décider si la solution que vous avez trouvée est privilégiée par rapport à toutes les autres. Une autre solution possible, qui contient deux constantes cosmogoniques, consiste à prendre les 10 anciennes équations $R_{ij} = 0$, qui ont encore ici un caractère invariant et à leur ajouter 12 équations de la forme

$$\Lambda^\mu_{\alpha\beta;\mu} = \phi_{\alpha,\beta} - \phi_{\beta,\alpha} = C\,h\,S^{\gamma\delta\mu}\phi_\mu, \tag{1}$$

$$(h\,S^{\alpha\beta\mu})_{,\mu} = C'h\,S^{\alpha\beta\mu}\phi_\mu; \tag{2}$$

dans les équations (1) les indices $\alpha\beta\gamma\delta$ sont fixes et forment une permutation paire des indices 1, 2, 3, 4. C et C' sont des constantes numériques quelconques [3].

La Physique correspondante serait *irréversible*, du moins si $C \neq 0$; en changeant l'orientation du 4-Bein [4], c'est-à-dire le sens du temps, les lois de la Physique cesseraient d'être vraies, mais cela ne se manifesterait pas en première approximation. Dans votre système, les 12 équations (1) et (2) figurent aussi, la constante C ayant la valeur 0 et le constante C' la valeur 1. Si $C = 0$ les équations (1) et (2) peuvent s'intégrer en introduisant deux fonctions ψ et χ

$$\phi_\alpha = -\frac{1}{C'\psi}\frac{\partial\psi}{\partial x_\alpha},$$

$$hS^{\alpha\beta\gamma} = \frac{1}{\psi}\frac{\partial\chi}{\partial x_\delta}.$$

L'existence de ces deux fonctions universelles définies à deux constantes près (ψ peut être remplacé par $a\psi$ et χ par $a\chi + b$) est assez curieuse. Dans la vieille Physique, il existait effectivement deux fonctions de cette nature, le potentiel de gravitation et le temps; comme ψ, le potentiel de gravitation est réversible et comme χ le temps est irréversible (change de signe avec l'orientation de l'espace-temps). Si

---

3. Si $\Lambda^\alpha_{\beta\lambda}$ sont les composantes du tenseur de torsion

$$\Phi_\alpha = \Lambda^\beta_{\alpha\beta}$$

et

$$S^\alpha_{\mu\nu} = \Lambda^\alpha_{\mu\nu} + \Lambda^\nu_{\alpha\mu} + \Lambda^\mu_{\nu\alpha}$$

Le système de 22 équations que propose ici E. Cartan, ainsi que celui à 16 équations qui sont des variantes de la solution d'Einstein sont exposés et discutés dans [13].

taking the 10 old equations $R_{ij} = 0$ that still have here an invariant character and adding to them 12 equations of the form

$$\Lambda^{\mu}_{\alpha\beta;\mu} = \phi_{\alpha,\beta} - \phi_{\beta,\alpha} = C\,h\,S^{\gamma\delta\mu}\phi_{\mu}, \tag{1}$$

$$(h\,S^{\alpha\beta\mu})_{,\mu} = C'h\,S^{\alpha\beta\mu}\phi_{\mu}; \tag{2}$$

In equations (1) the indices $\alpha\beta\gamma\delta$ are fixed and form an even permutation of the indices 1, 2, 3, 4. C and C′ are any numerical constants [3].

The corresponding physics would be irreversible, at least if $C \neq 0$; by changing the orientation of the 4-Bein [4], that is the direction of time, the laws of physics would cease to be true, but it would not show up in the first approximation. The 12 equations (1) and (2) also figure, in your system, the constant C having the value 0 and the constant C′ the value 1. If $C = 0$, the equations (1) and (2) can be integrated by introducing two functions ψ and χ

$$\phi_{\alpha} = -\frac{1}{C'\psi}\frac{\partial\psi}{\partial x_{\alpha}},$$

$$hS^{\alpha\beta\gamma} = \frac{1}{\psi}\frac{\partial\chi}{\partial x_{\delta}}.$$

The existence of these two universal functions defined up to two constants (ψ may be replaced by $a\psi$ and χ by $a\chi + b$) is rather strange. In the old physics, there did indeed exist two functions of that kind, the gravitational potential and time; like ψ the gravitational potential is reversible, and like χ time is irreversible (it changes sign with the orientation of space-time). If $C \neq 0$, two more universal functions are brought in, defined up to the transformations of a two-parameter group.

I was not able to completely solve the problem of determining if there are systems of 22 equations other than yours and the one I just indicated since I would have to set to zero a certain form cubic in the

---

4. Cartan utilise dans ses publications le terme de repère (cartésien, projectif, ...). Dans cette correspondance il emploie systématiquement 4-Bein, *n*-Bein; termes utilisés par Einstein dans ses notes aux *Sitzungsberichte*. Dans [21], en français, on trouve *n*-podes. Nous avons renoncé à traduire et nous n'avons pas employé d'italiques pour le signaler.

C $\neq$ 0, il s'introduit encore deux fonctions universelles, définies aux transformations près d'un groupe à deux paramètres.

(Je n'ai pu résoudre complètement le problème de savoir s'il existe d'autres systèmes de 22 équations que le vôtre et celui que je viens d'indiquer); il me resterait à identifier à zéro une certaine forme cubique des $\Lambda_{\alpha\beta}^{\gamma}$ dépendant de 16 paramètres! La partie des équations qui contient les dérivées $\Lambda_{\alpha\beta;\mu}^{\gamma}$ est bien déterminée, mais ce sont les termes complémentaires qui, même en les supposant simplement quadratiques, introduisent toutes les complications — et je suis toujours étonné que vous soyez arrivé à trouver vos 22 équations!!

Il y a d'autres possibilités, qui fournissent des schémas géométriques plus riches, tout en étant déterministes. On peut d'abord prendre un système de 15 équations: 6 d'entre elles expriment que le vecteur

$$a\phi_1 + bhS^{234}, \ a\phi_2 + bhS^{143}, \ a\phi_3 + bhS^{124}, \ a\phi_4 + bhS^{132}$$

(où a et b ($b \neq 0$) sont des constantes numériques), est le gradient d'une fonction universelle; les 9 autres sont de la forme

$$\Lambda_{\underline{\alpha}\mu;\mu}^{\beta} + \Lambda_{\underline{\beta}\mu;\mu}^{\alpha} + c(\phi_{\underline{\alpha},\beta} + \phi_{\underline{\beta},\alpha}) + d(hS^{\gamma\delta\mu})_{;\mu} - \frac{1}{2} g^{\alpha\beta} \left[\Lambda_{\underline{\rho}\mu;\mu}^{\rho} + c\phi_{\underline{\rho},\rho} \right] =$$

$$= A^{\alpha\beta} - \frac{1}{4} g^{\alpha\beta} A^{\mu\mu},$$

c et d étant des constantes numériques arbitraires, $A^{\alpha\beta}$ des fonctions des $\Lambda_{\mu\nu}^{\rho}$ *assujetties à l'unique condition de former un tenseur symétrique.* Si l'on prend pour $A^{\alpha\beta}$ des formes quadratiques, un tel tenseur dépend linéairement de 10 constantes numériques arbitraires. La Physique correspondante est réversible si la constante a est nulle et si la constante d est aussi nulle; il faut aussi dans ce cas limiter le tenseur $A^{\alpha\beta}$ qui ne doit plus dépendre que de 8 constantes arbitraires.

Enfin *peut-être* existe-t-il aussi des solutions formées de 16 équations; mais l'étude de ce cas conduit à des calculs aussi compliqués que celui de 22 équations et je n'ai pas eu le bonheur de tomber sur un système possible. La forme générale des équations d'un tel système s'obtiendrait en prenant d'abord 9 équations de la forme (3` indiquée ci-dessus puis une équation de la forme

$$\phi_{\underline{\mu};\mu} = C,$$

$A^\gamma_{\alpha\beta}$ and depending on 16 parameters! The part of the equations containing the derivatives $A^\gamma_{\alpha\beta;\mu}$ is well defined, but it is the complementary terms that, even when one simply assumes them to be quadratic, introduce all the complications — and it still astonishes me that you managed to find your 22 equations!

There are other possibilities giving rise to richer geometrical schema while remaining deterministic. First, one can take a system of 15 equations: 6 of them express the fact that the vector

$$a\phi_1 + bhS^{234}, \ a\phi_2 + bhS^{143}, \ a\phi_3 + bhS^{124}, \ a\phi_4 + bhS^{132},$$

(where a and b ($b \neq 0$) are numerical constants), is the gradient of a universal function; the other 9 are of the form

$$A^\beta_{\underline{\alpha}\mu;\mu} + A^\alpha_{\underline{\beta}\mu;\mu} + c(\phi_{\alpha,\beta} + \phi_{\beta,\underline{\alpha}}) + d(hS^{\gamma\delta\mu})_{;\mu} - \frac{1}{2} g^{\alpha\beta} [A^\rho_{\underline{\rho}\mu;\mu} + c\phi_{\rho,\rho}] =$$

$$= A^{\alpha\beta} - \frac{1}{4} g^{\alpha\beta} A^{\mu\mu},$$

c and d being arbitrary numerical constants and $A^{\alpha\beta}$ being functions of the $A^\rho_{\mu\nu}$ subject to the single condition that they form a symmetric tensor. If one substitutes quadratic forms for $A^{\alpha\beta}$, such a tensor depends linearly on 10 arbitrary numerical constants. The corresponding physics is reversible if the constant a is zero and if the constant d is also zero. In this case also one must restrict the tensor $A^{\alpha\beta}$ that now must depend on only 8 arbitrary constants.

Finally, *maybe* there are also solutions with 16 equations; but the study of this case leads to calculations as complicated as in the case of 22 equations, and I was not fortunate enough to come across a possible system. The general form for the equations of such a system would be obtained by first taking 9 equations of the form (3) as indicated above, then one equation of the form

$$\phi_{\underline{\mu};\mu} = C,$$

C being a scalar tensor (with 5 possible constants); and finally six equations such as

$$(hS^{34\mu})_{,\mu} + a(\phi_{1,2} + \phi_{2,1}) + bh(\phi_{3,\underline{4}} + \phi_{4,\underline{3}}) = B_{12}, \qquad (4)$$

27

C étant un tenseur scalaire (avec 5 constantes possibles); enfin six équations telles que

$$(hS^{34\mu})_{,\mu} + a(\phi_{1,2} + \phi_{2,1}) + bh(\phi_{\underline{3},\underline{4}} + \phi_{\underline{4},\underline{3}}) = B_{12}, \qquad (4)$$

où $B_{ij} = -B_{ji}$ est un tenseur antisymétrique. Si a ou d est différent de zéro, la Physique serait irréversible, et l'irréversibilité se manifesterait en première approximation.

J'ai dit plus haut que les solutions de 15 équations et celles (si elles existent) de 16 équations ont un contenu plus riche que celles de 22 équations; la solution générale des systèmes à 22 équations dépend en effet de 12 fonctions arbitraires de 3 variables, celle des systèmes de 16 équations dépend de 16 fonctions arbitraires et celle des systèmes de 15 équations, de 18 fonctions arbitraires. Pour le sens précis à donner à ces énoncés, vous serez obligé de vous reporter à la note technique ci-incluse. Il me semble après réflexion que le degré de généralité [5] du schéma géométrique correspondant à votre système de 22 équations est un peu faible, les anciennes théories classiques de la gravitation et de l'électromagnétisme donnant à la Physique un degré de généralité plus considérable. Je serai très heureux d'avoir votre opinion là-dessus.

Enfin vous verrez dans la note que le nombre des identités qui doivent exister entre les dérivées des premiers membres d'un système possible de n équations pour que ce système soit *non seulement compatible*, mais encore soit *en involution*, est *au moins égal* à $n - 12$, mais doit dans certains cas lui être *supérieur*: c'est du reste ce qui se passe pour votre système de 22 équations: les dérivées des premiers membres sont en réalité liées par $12 = n - 12 + 2$ relations linéaires identiques; ce n'est qu'en le modifiant par l'introduction d'une fonction auxiliaire ψ que vous êtes arrivé à retrouver votre nombre a priori de $n - 12$ identités. Bien entendu si l'on voulait simplement que le

---

5. La notion appelée ici *degré de généralité* sera invoquée très souvent dans la suite de la correspondance. Elle sera désignée de plusieurs façons différentes.

Voici un relevé, sous la plume de Cartan des diverses expressions utilisées avec référence aux lettres, notes et ouvrages où elles apparaissent:

degré de généralité: VII, XV, XXIV, XXXII, XXXVII, [13] et E. Cartan: Les systèmes différentiels extérieurs et leurs applications géométriques, Hermann, Paris, 1945.
indice de généralité: VII N, IX, XV, XX, XXI, XXII, XXIV, XXVII
degré d'arbitraire: XXXVII et [13].

where $B_{ij} = -B_{ji}$ is a skew-symmetric tensor. If a or d is different from zero, the physics would be irreversible and the irreversibility would show up in the first approximation.

I said before that the solutions of 15 equations and those (if they exist) of 16 equations have a richer content than that of 22 equations; in particular the general solution of the systems with 22 equations depends on 12 arbitrary functions of 3 variables; for a system with 16 equations, the general solution depends on 16 arbitrary functions and for a system with 15 equations on 18 arbitrary functions. For the precise meaning of these statements, you will have to refer to the enclosed technical note.

When thinking it over, I believe that the degree of generality [5] of the geometrical scheme corresponding to your system of 22 equations is a bit weak, the old classical theories of gravitation and electromagnetism give to physics a greater degree of generality. I would be very glad to have your opinion about this.

Finally, you will see in the note that the number of identities that must exist between the derivatives of the left-hand sides of a possible system of n equations, in order that that system be *not only compatible* but also in *involution*, is *at least equal* to $n - 12$, and in certain cases, must be larger. It is, in fact, what is going on in your system of 22 equations; the derivatives of the left hand sides are actually related by $12 = n - 12 + 2$ linear identities. It is only when modifying it by the introduction of an auxiliary function ψ, that you manage to recover your *a priori* number of $n - 12$ identities. Of course, if one simply wanted the system to be compatible, this number of identities would no longer be necessary, as can be shown by example.

I hope that you had a pleasant return voyage and that your health did not suffer too much from your trip to Paris. In any case, we

---

degré de détermination: XI, XV
degré d'indétermination: XXIV, XXXII.
    Einstein pour sa part utilise
Determination-grad: VIII, XVI, XXV
Index de généralité: (*Allgemeinheits-grad*): XVII, XXII, XXV, XXVII, XXVIII, XXIX, XXXI. Einstein dans XVIII la désigne par la lettre $\mathscr{I}$, notation utilisée souvent par les deux correspondants.
Degré d'arbitraire: XXXV (à la suite de [13]), XXXVI.

système soit compatible, il n'y aurait plus a priori nécessité d'un pareil nombre d'identités comme on pourrait le faire voir sur des exemples.

J'espère que vous avez fait un bon retour et que votre santé n'a pas eu trop à souffrir de votre voyage à Paris. En tous cas nous avons été tous très heureux que vous nous ayez initiés à vos plus récentes recherches. Je vous remercie encore de la confiance que vous avez témoignée dans mes qualités de mathématicien, confiance que je voudrais bien mériter, au moins a posteriori.

Veuillez agréer, cher et illustre Maître, l'expression de mon plus cordial dévouement.

E. Cartan

J'ai reçu il y a quelques jours les épreuves de ma note des *Math. Annalen*; je les ai renvoyées après correction et sans faire de modifications au texte, bien que certaines remarques que j'y fais n'aient plus maintenant aucune raison d'être [6].

Après avoir rédigé ma petite note, je m'aperçois qu'elle est bien longue; puisse-t-elle au moins être claire!

---

6. Cartan fait sans doute allusion à ses remarques du paragraphe 9 de [9] où il discute des possibilités offertes par les différentes composantes irréductibles du

were all very pleased that you introduced us to your latest research. I thank you again for the trust you put in my abilities as a mathematician, a trust that I should like to deserve, at least, *a posteriori*

E. Cartan

A few days ago, I received, the proofs of my article for the *Math. Annalen*; I returned them corrected but without modifying the text, although certain remarks I make in it have now lost their raison d'être [6].

After having written my little note, I realized that it is quite long: may it at least be clear!

---

tenseur de torsion, discussion superflue depuis qu'Einstein a choisi ses équations de champ.

VII N

Note jointe à la lettre du 3 décembre 1929.

I. Les systèmes d'équations aux dérivées partielles qui respectent le déterminisme

Prenons une théorie physique telle que les grandeurs physiques qui interviennent dans cette théorie soient liées par un certain système d'équations aux dérivées partielles. D'après le langage vulgaire, cette théorie respecte le déterminisme si, étant donné à un instant t les valeurs numériques des grandeurs physiques aux différents points de l'espace, les valeurs de ces grandeurs sont déterminées à tout instant de la durée.

Prenons par exemple la vieille théorie de la gravitation et de la matière, où l'on fait complètement abstraction de l'électromagnétisme, et qui est régie par les équations

$$\frac{\partial Y}{\partial z} - \frac{\partial Z}{\partial y} = 0, \ \frac{\partial Z}{\partial x} - \frac{\partial X}{\partial z} = 0, \ \frac{\partial X}{\partial y} - \frac{\partial Y}{\partial x} = 0$$

$$\frac{\partial X}{\partial x} + \frac{\partial Y}{\partial y} + \frac{\partial Z}{\partial z} = -4\pi f \rho,$$

$$\frac{\partial \rho}{\partial t} + \frac{\partial(\rho u)}{\partial x} + \frac{\partial(\rho v)}{\partial y} + \frac{\partial(\rho w)}{\partial z} = 0,$$

$$\frac{\partial u}{\partial t} + u\frac{\partial u}{\partial x} + v\frac{\partial u}{\partial y} + w\frac{\partial u}{\partial z} = X,$$

$$\frac{\partial v}{\partial t} + u\frac{\partial v}{\partial x} + v\frac{\partial v}{\partial y} + w\frac{\partial v}{\partial z} = Y,$$

$$\frac{\partial w}{\partial t} + u\frac{\partial w}{\partial x} + v\frac{\partial w}{\partial y} + w\frac{\partial w}{\partial z} = Z;$$

$$(1)$$

il entre 7 grandeurs physiques, les composantes X, Y, Z de l'accélé-ration de gravitation, la densité de matière ρ et les composantes u, v, w de la vitesse d'un élément de matière. Nous supposons la matière continue, mais dénuée de toute tension ou pression.

3 December 1929

A VII N

I. Systems of partial differential equations
that respect determinism

Let us consider a physical theory whose physical quantities are related by some system of partial differential equations. In common parlance, this theory respects determinism if, at an instant t the numerical values of the physical quantities at the different points of space are given, the values of these quantities are determined at any other instant of time.

Let us take, for example, the old theory of matter and gravitation, where one disregards electromagnetism completely and that is governed by the equations

$$\frac{\partial Y}{\partial z} - \frac{\partial Z}{\partial y} = 0, \; \frac{\partial Z}{\partial x} - \frac{\partial X}{\partial z} = 0, \; \frac{\partial X}{\partial y} - \frac{\partial Y}{\partial x} = 0$$

$$\frac{\partial X}{\partial x} + \frac{\partial Y}{\partial y} + \frac{\partial Z}{\partial z} = -4\pi f\rho,$$

$$\frac{\partial \rho}{\partial t} + \frac{\partial(\rho u)}{\partial x} + \frac{\partial(\rho v)}{\partial y} + \frac{\partial(\rho w)}{\partial z} = 0,$$

$$\frac{\partial u}{\partial t} + u\frac{\partial u}{\partial x} + v\frac{\partial u}{\partial y} + w\frac{\partial u}{\partial z} = X,$$

$$\frac{\partial v}{\partial t} + u\frac{\partial v}{\partial x} + v\frac{\partial v}{\partial y} + w\frac{\partial v}{\partial z} = Y,$$

$$\frac{\partial w}{\partial t} + u\frac{\partial w}{\partial x} + v\frac{\partial w}{\partial y} + w\frac{\partial w}{\partial z} = Z;$$

$$(1)$$

7 physical quantities enter the equations, the components X, Y, Z of the gravitational acceleration, the matter density ρ and the components u, v, w of the velocity of an element of matter. We assume the matter to be continuous but free from stress or pressure.

33

Au sens vulgaire du terme, les équations (1) ne garantissent pas le déterminisme, car la connaissance à un instant t de X, Y, Z, ρ, u, v, w ne permet pas de calculer à cet instant toutes les dérivées $\frac{\partial X}{\partial t}$, etc. Le déterminisme est sauvé dans la vieille théorie en admettant que la force de gravitation est complètement déterminée à chaque instant, suivant la loi de Newton, par l'état de la matière dans tout l'Univers à cet instant. Mais c'est là en quelque sorte un déterminisme *intégral* (opposé à *local*) qui exige la connaissance intégrale de l'état de l'Univers à un instant t pour qu'on en puisse déduire, même localement, l'état de l'Univers à un instant voisin.

Le système (1) garantit néanmoins le déterminisme *local* si l'on conçoit le déterminisme d'une manière un peu plus générale. En effet, supposons connues les grandeurs physiques X, Y, Z, ρ, u, v, w dans une section à 3 dimensions de l'espace-temps définie par exemple par

$$t = \phi(x,y,z).$$

Il est facile alors de voir que si la fonction ϕ n'est pas trop particulière, la connaissance des 7 grandeurs physiques dans *une portion* de la section à 3 dimensions considérée suffit pour entraîner la connaissance de ces 7 grandeurs dans une portion correspondante d'une section infiniment voisine. En ce sens, il y a déterminisme *local,* la connaissance locale de l'état de l'Univers dans une section à 3 dimensions de l'espace-temps suffisant pour entraîner toute l'évolution de l'Univers dans le voisinage où on se trouve placé. Les seules sections à 3 dimensions de l'espace-temps qui fassent exception sont: 1° Les sections $t = C^{te}$; 2° les sections qu'on obtient en prenant à un instant donné une surface S, considérant les molécules qui se trouvent à cet instant sur cette surface et suivant ces mêmes molécules dans toute la durée. Les deux catégories de variétés à 3 dimensions ainsi obtenues sont ce que les mathématiciens appellent les variétés *caractéristiques* du système (1).

Il semble donc naturel d'appeler déterministe une théorie physique régie par des équations aux dérivées partielles telles que la connaissance des grandeurs physiques dans une section *arbitraire* à 3 dimensions de l'espace-temps détermine complètement ces grandeurs dans tout l'espace-temps. En ce sens les équations (1) sont déterministes. C'est aussi dans ce sens que les équations susceptibles de fournir une

In the common meaning of the term, the equations (1) do not guarantee determinism, since a knowledge of X, Y, Z, ρ, u, v, w at an instant t does not allow one to calculate, at that instant, all the derivatives $\frac{\partial X}{\partial t}$, etc. Determinism is preserved in the old theory because one assumes that the gravitational force is completely determined at any moment according to Newton's law, by the state of the matter in the whole Universe at that moment. But this is, in a way, a *global* determinism (as opposed to *local*) for which one needs a complete knowledge of the state of the Universe at an instant t in order to be able to deduce from it, even locally, the state of the Universe at a neighbouring instant.

The system (1) nevertheless guarantees local determinism if one conceives determinism in a slightly more general way. Let us assume that the physical quantities X, Y, Z, ρ, u, v, w are known in a 3-dimensional cross-section of space-time defined by

$$t = \phi(x,y,z).$$

It is easy to see that, if the function ϕ is not too special, the knowledge of the 7 physical quantities in *a domain* of the 3-dimensional section is sufficient to entail a knowledge of these 7 quantities in a corresponding domain of an infinitesimally close section. In this sense, there is *local* determinism, a local knowledge of the state of the Universe in a 3-dimensional section of space-time being sufficient to entail the whole evolution of the Universe in its neighbourhood. The only 3-dimensional sections of space-time that are exceptions are: 1) the sections $t = C^{onst}$, 2) the sections obtained by taking a surface S at a given moment, looking at the molecules lying at that moment on this surface and following these same molecules for all time. The two types of 3-dimensional manifolds obtained in this way are what mathematicians call the *characteristic* manifolds of the system (1).

It seem natural therefore, to call deterministic a physical theory governed by partial differential equations such that a knowledge of the physical quantities on an *arbitrary* 3-dimensional section of space-time determines these quantities completely in the whole space-time. In this sense equations (1) are deterministic. It is also in this sense that the equations likely to give a unified theory of gravitation and electromagnetism must be deterministic.

théorie unitaire de la gravitation et de l'électromagnétisme doivent être déterministes.

II. Les systèmes en involution

Considérons un système d'équations aux dérivées partielles linéaires (par rapport aux dérivées des fonctions inconnues) du premier ordre, par exemple le système (1). Introduisons d'abord quelques notions préliminaires.

Appelons *solution à une dimension* du système une suite continue de valeurs des fonctions inconnues X, ..., w données sur une ligne (à 1 dimension) de l'espace-temps; cette suite de valeurs doit satisfaire à la condition que les valeurs des dérivées des fonctions dans la direction de la ligne soient compatibles avec au moins un choix des dérivées dans trois autres directions indépendantes de l'espace-temps, de manière à satisfaire aux équations (1). Par exemple, dans le cas des équations (1), on a une solution à 1 dimension en se donnant arbitrairement pour $y = z = t = 0$, les grandeurs X, Y, Z, ρ, u, v, w en fonction de x.

Appelons *solution à deux dimensions* une suite de valeurs des fonctions données en tous les points d'une surface, avec la condition que les dérivées de ces fonctions dans deux directions indépendantes de la surface soient compatibles avec au moins un choix des dérivées dans deux autres directions, de manière à satisfaire aux équations (1). Par exemple on pourra se donner sur la surface $z = t = 0$ les fonctions X, Y, Z, ρ, u, v, w de x, y, mais à la condition que $\frac{\partial X}{\partial y} - \frac{\partial Y}{\partial x}$ soit nul, c'est-à-dire que $X dx + Y dy$ soit, sur la surface, une différentielle exacte.

On définit de même une solution à 3 dimensions et aussi, par analogie, une solution à 0 dimension, qui sera constituée par un système de valeurs numériques attribuées, en un point donné, aux différentes fonctions inconnues.

Cela posé, le système différentiel sera dit *en involution* si toute solution à 0 dimension fait partie d'au moins une solution à 1 dimension, si toute solution *arbitraire* à 1 dimension fait partie d'au moins une solution à 2 dimensions et ainsi de suite.

On peut énoncer de la manière suivante les conditions nécessaires et suffisantes pour que le système soit en involution:

II. SYSTEMS IN INVOLUTION

Consider a system of partial differential equations which are linear (with respect to the derivatives of the unknown functions) and of first order, for example the system (1). Let us first introduce a few preliminary notions.

Let us call a *one-dimensional solution* of the system a continuous sequence of values of the unknown functions X, ..., w given on a line (of dimension 1) of space-time; this sequence of values must satisfy the condition that the values of the derivatives of the unknown functions in the direction of the line should be compatible with at least one choice of the derivatives in three other independent directions of space-time, so as to satisfy equations (1). For example, in the case of equations (1), one has a 1-dimensional solution by taking arbitrary functions of x for X, Y, Z ρ, u, v, w at $y = z = t = 0$.

Let us call a 2-*dimensional solution* a sequence of values of the unknown functions given at every point of a surface, with the condition that the derivatives of these functions in two independent directions of the surface should be compatible with one choice of the derivatives in two other directions, so as to satisfy equations (1). For instance, on the surface $z = t = 0$, X, Y, Z, ρ, u, v, w may be given as functions of x and y but with the condition that $\dfrac{\partial X}{\partial y} - \dfrac{\partial Y}{\partial x} = 0$, i.e. that $Xdx + Ydy$ should be an exact differential on the surface.

One defines in the same way a 3-dimensional solution and also, by analogy, a zero-dimensional solution, that will consist of a system of numerical values assigned, at a given point to the different unknown functions.

This being granted, the differential system will be said to be *in involution* if any 0-dimensional solution is part of at least one 1-dimensional solution, if any *arbitrary* 1-dimensional solution is part of at least one 2-dimensional solution, and so on.

The necessary and sufficient conditions for the system to be in involution can be stated as follows:

Take at any point of space an arbitrary direction 1 *and let* r_1 *be the number of linear relations that hold in the given system among the p derivatives of the unknown functions in the direction* 1;

Prenons en un point quelconque de l'espace une direction arbitraire 1, *et soit* r_1 *le nombre des relations linéaires que le système donné établit entre les p dérivées des fonctions inconnues suivant la direction* 1;

prenons de même une nouvelle direction 2, *et soit* $r_2 \geqslant r_1$ *le nombre des relations linéaires que le système donné établit entre les 2p dérivées des fonctions inconnues suivant les directions* 1 *et* 2;

prenons enfin une troisième direction arbitraire 3 *indépendante des deux premières et soit* r_3 *le nomvre des relations linéaires que le système établit entre les 3p dérivées prises suivant les* 3 *directions* 1, 2, 3.

Pour que le système soit en involution il faut et il suffit qu'il existe, entre les 4n dérivées des premiers membres des équations, $r_1 + r_2 + r_3$ *identités (en tenant compte des équations du système).*

On démontre du reste que dans aucun cas le nombre des identités possibles ne peut dépasser $r_1 + r_2 + r_3$.

Le système en involution sera de plus déterministe si l'on a

$$r_3 = n - p.$$

Les conditions pour qu'un système soit en involution et déterministe sont donc que r_3 *soit égal à* $n - p$ *et qu'il existe* $n - p + r_1 + r_2$ *relations linéaires identiques entre les 4n dérivées des premiers membres.*

REMARQUE. — Les entiers r_1, r_2, r_3 se rapportent à un choix arbitraire des directions 1, 2, 3; il se peut par exemple que pour un choix particulier des directions 1 et 2, on obtienne plus de r_2 relations linéaires entre les dérivées prises suivant ces deux directions. Si cette circonstance se produit pour tous les points d'une surface de l'espace-temps, les directions 1 et 2 étant tangentes à la surface, cette surface est une *caractéristique*. Il peut donc y avoir des variétés caractéristiques à 2 dimensions (et même à une) aussi bien qu'à 3 dimensions.

EXEMPLE. — Reprenons le système (1). La direction la plus générale peut être supposée, par un choix convenable des axes, définie par

$$\frac{dx}{1} = \frac{dy}{0} = \frac{dz}{0} = \frac{dt}{\alpha};$$

on trouve alors $r_1 = 0$. Un système arbitraire de deux directions peut

take, in the same way, a new direction 2, and let $r_2 \geqslant r_1$ be the number of linear relations that hold in the given system among the 2p derivatives of the unknown functions in the directions 1 and 2;

finally, take a third arbitrary direction independent of the other two and let r_3 be the number of linear relations that hold in the system among the 3p derivatives in the 3 directions 1, 2, 3.

The system is in involution if and only if there exist $r_1 + r_2 + r_3$ identities among the 4n derivatives of the left hand sides of the equations (taking into account the equations of the system).

Moreover, it can be proven that the number of possible identities can never be larger than $r_1 + r_2 + r_3$.

The system in involution will, in addition, be deterministic if one has

$$r_3 = n - p.$$

Therefore, the conditions for a system to be in involution and deterministic are that r_3 should be equal to $n - p$ and that there should exist $n - p + r_1 + r_2$ linear identities among the 4n derivatives of the left hand sides.

REMARK. — The integers r_1, r_2, r_3 refer to an arbitrary choice of the directions 1, 2, 3; it may happen, for example, that for a particular choice of the directions 1 and 2, one obtains more than r_2 linear relations among the derivatives taken in these directions. If this happens for all the points of a surface in space-time, the directions 1 and 2 being tangent to the surface, this surface is a *characteristic*. Hence, 2-dimensional characteristic manifolds can exist (even 1-dimensional ones) as well as 3-dimensional ones.

EXAMPLE: Let us again take system (1). With a suitable choice of the axes, the most general direction can be assumed to be defined by

$$\frac{dx}{1} = \frac{dy}{0} = \frac{dz}{0} = \frac{dt}{\alpha};$$

then one has $r_1 = 0$. A system of two arbitrary directions is obtained

être obtenu en adjoignant à la direction précédente la direction

$$\frac{dx}{0} = \frac{dy}{1} = \frac{dz}{0} = \frac{dt}{\beta};$$

on trouve alors $r_2 = 0$, à moins que l'on n'ait $\alpha = \beta = 0$ (les caractéristiques à deux dimensions sont des surfaces de l'espace prises à un instant fixe). Enfin on trouve $r_3 = 1 = n - p = 8 - 7$. Le système est en involution car les dérivées des premiers membres sont liées par une identité ($1 = r_1 + r_2 + r_3$), à savoir

$$\frac{\partial}{\partial x}\left(\frac{\partial Y}{\partial z} - \frac{\partial Z}{\partial y}\right) + \frac{\partial}{\partial y}\left(\frac{\partial Z}{\partial x} - \frac{\partial X}{\partial z}\right) + \frac{\partial}{\partial z}\left(\frac{\partial X}{\partial y} - \frac{\partial Y}{\partial x}\right) = 0.$$

III. Indice de généralité d'un système déterministe en involution

Prenons un système en involution *non déterministe*. Par exemple, en supposant que les variables indépendantes x_1, x_2, x_3, x_4 ne soient pas choisies d'une manière trop particulière, les équations du système établissent entre les dérivées par rapport à x_1, x_2, x_3 un nombre r_3 de relations *supérieur* à $n - p$. Il en résulte que les équations peuvent être résolues par rapport à $n - r_3$ seulement des dérivées prises par rapport à x_4, par exemple les dérivées de $n - r_3 < p$ premières fonctions inconnues. On pourra alors prendre pour les $p - (n - r_3)$ dernières fonctions inconnues des fonctions *arbitraires* de x_1, x_2, x_3, x_4. Le système qui lie les $n - r_3$ autres fonctions inconnues est encore en involution et est devenu déterministe. On peut dire que la solution générale du système donné dépend de $r_3 + p - n$ fonctions *arbitraires* de quatre variables.

Prenons maintenant un système déterministe. Toutes solution du système est parfaitement déterminée par la solution à 3 dimensions obtenue en faisant une section à 3 dimensions *déterminée* (non caractéristique) par exemple $x_4 = 0$ de l'espace temps. Les solutions à 3 dimensions sont définies par le système obtenu en éliminant les dérivées par rapport à x_4 des équations données; ce nouveau système contient $n - p$ équations, p fonctions inconnues, et il est encore en involution avec les mêmes valeurs de r_1 et de r_2. Par suite la solution générale dépend de

$$r_2 + p - (n - p) = r_2 + 2p - n$$

fonctions *arbitraires* de 3 variables.

by adding to the previous one, the direction

$$\frac{dx}{0} = \frac{dy}{1} = \frac{dz}{0} = \frac{dt}{\beta};$$

then one has $r_2 = 0$, unless $\alpha = \beta = 0$ (the 2-dimensional characteristics are surfaces of constant time). Finally, one obtains $r_3 = 1 = n - p = 8 - 7$. The system is in involution since the derivatives of the left-hand sides are related by one identity ($1 = r_1 + r_2 + r_3$) namely

$$\frac{\partial}{\partial x}\left(\frac{\partial Y}{\partial z} - \frac{\partial Z}{\partial y}\right) + \frac{\partial}{\partial y}\left(\frac{\partial Z}{\partial x} - \frac{\partial X}{\partial z}\right) + \frac{\partial}{\partial z}\left(\frac{\partial X}{\partial y} - \frac{\partial Y}{\partial x}\right) = 0.$$

III. GENERALITY INDEX OF A DETERMINISTIC SYSTEM IN INVOLUTION

Let us take a *non-deterministic* system in involution. For instance, since we assume that the independent variables x_1, x_2, x_3, x_4 are not chosen in too special a way, the equations of the system establish a number r_3 of relations *larger* than $n - p$ among the derivatives with respect to x_1, x_2, x_3. It then follows that the equations can be solved with respect to only $n - r_3$ of the x_4-derivatives, for example the derivatives of the first $n - r_3 < p$ unknown functions. The remaining $p - (n - r_3)$ unknown functions can then be chosen to be *arbitrary* functions of x_1, x_2, x_3, x_4. The system relating the $n - r_3$ other unknown functions is still in involution and has become deterministic. One may say that the general solution of the given system depends on $r_3 + p - n$ *arbitrary* functions of four variables.

Let us now take a deterministic system. Any solution of the system is perfectly determined by the 3-dimensional solution obtained by making a 3-dimensional *determined* section (non characteristic), for example $x_4 = 0$ in space-time. The 3-dimensional solutions are defined by the system obtained in eliminating the x_4-derivatives from the given equations. This new system contains $n - p$ equations and p unknown functions and is still in involution with the same values of r_1 and r_2. As a result, the general solution depends on

$$r_2 + p - (n - p) = r_2 + 2p - n$$

arbitrary functions of 3 variables.

On peut donc dire que la solution générale du système déterministe en involution donné dépend de $r_2 + 2p - n$ fonctions arbitraires de 3 variables, en ce sens que la solution à 3 dimensions ($x_4 = 0$) qui détermine la solution la plus générale peut être obtenue en se donnant arbitrairement $r_2 + 2p - n$ des fonctions inconnues en fonction de x_1, x_2, x_3.

L'entier $r_2 + 2p - n$ peut être appelé l'*indice de généralité* du système déterministe en involution considéré. On peut démontrer qu'il est toujours le même, de quelque manière qu'on choisisse les fonctions inconnues, même si on ajoute de nouvelles fonctions inconnues, dérivées jusqu'à un ordre quelconque des premières. *L'indice de généralité joue donc un rôle important.*

Si nous reprenons encore une fois le système (1), nous voyons que $r_2 = 0$, $p = 7$, $n = 8$, $r_2 + 2p - n = 6$. Par exemple toute solution du système (1) est déterminée si on connaît les grandeurs X, Y, Z, ρ, u, v, w dans la section $z = 0$ de l'espace-temps. Le système qui définit les solutions à 3 dimensions correspondant à cette section est

$$\frac{\partial X}{\partial y} - \frac{\partial Y}{\partial x} = 0;$$

on peut, pour obtenir sa solution générale, se donner arbitrairement Y, Z, ρ, u, v, w en fonction de x, y, t. *La théorie classique de la gravitation et de la matière continue (sans tension) fournit donc un système déterministe en involution dont l'indice de généralité est 6.*

Les équations de Maxwell jointes aux 4 équations qui donnent le mouvement de l'électricité (continue) constituent également un système en involution déterministe dont l'indice de généralité est 8. Si l'on prend en effet la section $t = 0$, les équations donnent, par élimination des dérivées par rapport à t,

$$\frac{\partial H_x}{\partial x} + \frac{\partial H_y}{\partial y} + \frac{\partial H_z}{\partial z} = 0,$$

$$\frac{\partial E_x}{\partial x} + \frac{\partial E_y}{\partial y} + \frac{\partial E_z}{\partial z} = \rho,$$

on a ici $n = 12$, $p = 10$ (champ électrique, champ magnétique, densité de charge, densité de courant), $r_2 = 0$, $r_2 + 2p - n = 8$; on peut choisir arbitrairement en fonction de x, y, z, pour $t = 0$, les quantités H_x, H_y, E_x, E_y, ρ, i_x, i_y, i_z.

So one can say that the general solution of a given deterministic system in involution depends on $r_2 + 2p - n$ arbitrary functions of 3 variables in the sense that the 3-dimensional solution ($x_4 = 0$) that determines the most general solution can be obtained by arbitrarily taking $r_2 + 2p - n$ of the unknown functions to be functions of x_1, x_2, x_3.

The integer $r_2 + 2p - n$ can be called the *generality index* of the considered deterministic system in involution. It can be shown that the index is always the same, whatever way the unknown functions have been chosen, and even if one adds new unknown functions that are derivatives up to any order of the first ones. *The generality index thus plays an important role.*

If we return to the system (1), we see that $r_2 = 0$, $p = 7$, $n = 8$, $r_2 + 2p - n = 6$. For example, any solution of system (1) is determined if the quantities X, Y, Z, ρ, u, v, w are known in the section $z = 0$ of space-time. The system defining the 3 dimensional solutions corresponding to this section is

$$\frac{\partial X}{\partial y} - \frac{\partial Y}{\partial x} = 0;$$

to get the general solution, one can arbitrarily give Y, Z, ρ, u, v, w as functions of x, y, t. *Hence the classical theory of gravitation and (stress-free) continuous matter provides a deterministic system in involution with a generality index equal to 6.*

The Maxwell's equations together with the 4 equations that give the motion of (continuous) electricity, also form a deterministic system in involution for which the generality index is 8. Indeed if one takes the section $t = 0$, and eliminate the t-derivatives the equations yield

$$\frac{\partial H_x}{\partial x} + \frac{\partial H_y}{\partial y} + \frac{\partial H_z}{\partial z} = 0,$$

$$\frac{\partial E_x}{\partial x} + \frac{\partial E_y}{\partial y} + \frac{\partial E_z}{\partial z} = \rho,$$

Here one has $n = 12$, $p = 10$ (electric field, magnetic field, charge density, current density), $r_2 = 0$, $r_2 + 2p - n = 8$; at $t = 0$, the quantities H_x, H_y, E_x, E_y, ρ, i_x, i_y, i_z can be chosen arbitrarily as functions of x, y, z.

IV. La théorie unitaire;
position mathématique du problème

Convenons d'abord d'une définition. Dans toute théorie physique il intervient un certain nombre de grandeurs physiques; ce nombre est jusqu'à un certain point indéterminé puisque les dérivées partielles d'une grandeur physique sont aussi des grandeurs physiques. *Nous dirons qu'un certain nombre de grandeurs physiques constituent un système fondamental si toute la théorie physique peut être condensée dans un système en involution d'équations aux dérivées partielles du premier ordre par rapport aux grandeurs physiques considérées.*

Bien entendu une même théorie physique peut comporter plusieurs systèmes fondamentaux de grandeurs physiques.

Cela posé, partons avec Einstein de la notion d'espace riemannien sans courbure, ou avec parallélisme absolu. Le problème consiste à restreindre cette notion générale par des conditions qui permettent d'obtenir une représentation adéquate de la gravitation et de l'électromagnétisme. Prenons un de ces espaces ainsi restreints, que nous supposerons *sans singularité; on dira y trouver le champ gravitationnel-électromagnétique, ainsi que la matière et l'électricité à l'état continu.*

« Les espaces qu'il s'agit de déterminer doivent donner toute la nature, mais pas plus que la nature. » Il en résulte que tout invariant différentiel d'un tel espace doit être une grandeur physique. Les premières grandeurs physiques qui se présentent au point de vue géométrique sont les composantes $\Lambda_{\alpha\beta}^{\gamma}$ de la torsion de l'espace. Nous admettons la première hypothèse suivante

Hypothèse I. — *Les composantes de la torsion constituent un système fondamental de grandeurs physiques.*

Nous pouvons maintenant nous placer à deux points de vue bien différents, celui du *physicien* qui vit dans un des Univers cherchés et celui du *démiurge* qui construit cet Univers.

I. *Point de vue du physicien.* — Le physicien a constaté l'existence dans son Univers d'une métrique et d'un parallélisme (au moins de proche en proche). En outre il est arrivé à établir le système d'équations aux dérivées partielles auxquelles obéissent les 24 grandeurs phy-

IV. THE UNIFIED THEORY: THE MATHEMATICAL SETTING OF THE PROBLEM

Let us first agree on a definition. In any physical theory a certain number of physical quantities are present; this number is to a certain extent undetermined since the partial derivatives of a physical quantity are also physical quantities. *We shall say that a certain number of physical quantities form a fundamental system if the entire physical theory can be condensed into a system in involution of first order partial differential equations with respect to the physical quantities under consideration.*

Of course, the same physical theory may allow for many fundamental systems of physical quantities.

This being granted, let us like Einstein start from the notion of a Riemannian space without curvature, i.e. with absolute parallelism. The problem lies in restricting this general concept by conditions which lead to an adequate representation of gravitation and electromagnetism. Let us take one of these spaces restricted in this way, one which we shall assume singularity-free; *the gravitational-electromagnetic field, as well as matter and electricity, will be assumed to be in a continuous state.*

" The spaces that are to be determined, must yield all of Nature, but no more than Nature. " It follows that any differential invariant of such a space must be a physical quantity. From the geometrical point of view the first physical quantities to appear are the components $\Lambda_{\alpha\beta}^{\gamma}$ of the torsion of the space. We make the following assumption:

ASSUMPTION I. — *The components of the torsion form a fundamental system of physical quantities.*

Now we can adopt two different points of view, the point of view of the *physicist* living in one of these Universes and that of the demiurge who builds this Universe.

I. *The physicist's point of view.*

The physicist has noticed the existence in his Universe of a metric and a parallelism (at least locally). Moreover, he has managed to establish the system of partial differential equations satisfied by the 24 fun-

siques fondamentales $\Lambda^{\gamma}_{\alpha\beta}$. D'après l'hypothèse I, ce système est en involution. Ajoutons

HYPOTHÈSE II. — *Le système différentiel auquel obéissent les 24 grandeurs physiques fondamentales est déterministe.*

Le système d'équations aux dérivées partielles auquel obéissent les 24 grandeurs physiques fondamentales contiendra évidemment les 16 équations

$$\Lambda^{\mu}_{\alpha\beta;\gamma} + \Lambda^{\mu}_{\beta\gamma;\alpha} + \Lambda^{\mu}_{\gamma\alpha;\beta} + \Lambda^{\rho}_{\alpha\beta}\Lambda^{\mu}_{\rho\gamma} + \Lambda^{\rho}_{\beta\gamma}\Lambda^{\mu}_{\rho\alpha} + \Lambda^{\rho}_{\gamma\alpha}\Lambda^{\mu}_{\rho\beta} = 0 \qquad (1)$$

qui, *pour le physicien,* ne sont pas des identités, mais expriment des lois physiques de la nature. Il s'y ajoutera une certain nombre n d'équations (2). D'après les hypothèses I et II, on aura les propriétés suivantes :

1° *les équations (1) et (2) contiennent 24 combinaisons linéaires indépendantes par rapport aux dérivées $\Lambda^{\gamma}_{\alpha\beta;4}$ (hypothèse déterministe) ;*

2° *si l'on peut tirer des équations (1) et (2) r_1 relations ne contenant que les dérivées $\Lambda^{\gamma}_{\alpha\beta;1}$ et r_2 relations ne contenant que les dérivées $\Lambda^{\gamma}_{\alpha\beta;1}$ et $\Lambda^{\gamma}_{\alpha\beta;2}$, les $64 + 4n$ dérivées des premiers membres des équations (1) et (2) doivent être liées par*

$$(16 + n) - 24 + r_1 + r_2 = n - 8 + r_1 + r_2$$

relations linéaires identiques.

Comme les 64 dérivées des 16 premiers membres des équations (1) sont déjà liées par 4 relations identiques, il en résulte que *les $4n$ dérivées des premiers membres des équations (2) doivent être liées par*

$$n - 8 + r_1 + r_2 - 4 = n - 12 + r_1 + r_2$$

relations identiques (compte tenu des équations (1) et de leurs dérivées).

On retrouve le nombre $n - 12$ fourni par un premier raisonnement a priori, si l'on a $r_1 = r_2 = 0$ (c'est-à-dire $r_2 = 0$).

Prenons par exemple pour équations (2) les 22 équations d'Einstein

$$\begin{cases} F_{\alpha\beta} \equiv \Lambda^{\mu}_{\alpha\beta;\mu} = 0, \\ G^{\alpha\beta} \equiv \Lambda^{\beta}_{\underline{\alpha}\mu;\mu} + \Lambda^{\rho}_{\underline{\alpha}\underline{\mu}}\Lambda^{\beta}_{\mu\rho} = 0. \end{cases} \qquad (2)$$

damental physical quantities $\Lambda_{\alpha\beta}^{\gamma}$. By virtue of assumption I, this system is in involution. Let us add

ASSUMPTION II. — *The differential system satisfied by the 24 fundamental physical quantities is deterministic.*

The system of partial differential equations satisfied by the 24 fundamental physical quantities will of course contain the 16 equations

$$\Lambda_{\alpha\beta;\gamma}^{\mu} + \Lambda_{\beta\gamma;\alpha}^{\mu} + \Lambda_{\gamma\alpha;\beta}^{\mu} + \Lambda_{\alpha\beta}^{\rho}\Lambda_{\rho\gamma}^{\mu} + \Lambda_{\beta\gamma}^{\rho}\Lambda_{\rho\alpha}^{\mu} + \Lambda_{\gamma\alpha}^{\rho}\Lambda_{\rho\beta}^{\mu} = 0 \qquad (1)$$

that, *for the physicist*, are not identities, but rather express physical laws of nature. To these we will add a certain number n of equations (2). According to assumptions I and II, the following properties will hold:

1) *equations (1) and (2) will contain 24 linearly independent combinations of the derivatives $\Lambda_{\alpha\beta;4}^{\gamma}$ (deterministic assumption);*

2) *if one can extract from equations (1) and (2) r_1 relations containing only the derivatives $\Lambda_{\alpha\beta;1}^{\gamma}$ and r_2 relations containing only $\Lambda_{\alpha\beta;1}^{\gamma}$ and $\Lambda_{\alpha\beta;2}^{\gamma}$, the $64 + 4n$ derivatives of the left hand sides of the equations (1) and (2) must be related by*

$$(16 + n) - 24 + r_1 + r_2 = n - 8 + r_1 + r_2$$

linear identities.

Since the 64 derivatives of the 16 left hand sides of equations (1) are already related by 4 identities, it follows that *the 4n derivatives of the left hand sides of equations (2) must be related by*

$$n - 8 + r_1 + r_2 - 4 = n - 12 + r_1 + r_2$$

identities (taking equations (1) and their derivatives into account).

One will recover the number $n - 12$ obtained by a first a priori reasoning, if one has $r_1 = r_2 = 0$ (i.e. $r_2 = 0$).

Let us take for instance, the 22 equations of Einstein, for our equations (2),

$$\begin{cases} F_{\alpha\beta} \equiv \Lambda_{\alpha\beta;\mu}^{\mu} = 0, \\ G^{\alpha\beta} \equiv \Lambda_{\underline{\alpha}\underline{\mu};\mu}^{\beta} + \Lambda_{\underline{\alpha}\underline{\mu}}^{\rho}\Lambda_{\mu\rho}^{\beta} = 0. \end{cases} \qquad (2)$$

47

On a ici $r_1 = 0$ et $r_2 = 2$, les 2 relations entre les $\varLambda^\gamma_{\alpha\beta;1}$ et $\varLambda^\gamma_{\alpha\beta;2}$ étant, en tenant compte des identités (1),

$$\varLambda^\mu_{12;\mu} = 0 \quad \text{et} \quad G^{34} - G^{43} - F^{34} = 0\ ^1.$$

Le système est en involution, car on a les $12 = n - 12 + 2$ identités [2]

$$G^{\alpha\mu}_{\ ;\mu} - F^{\alpha\mu}_{\ ;\mu} \equiv -\varLambda^\gamma_{\alpha\underline{\mu}}F_{\mu\nu},$$

$$F_{\alpha\beta;\gamma} + F_{\beta\gamma;\alpha} + F_{\gamma\alpha;\beta} \equiv \varLambda^\rho_{\alpha\beta}F_{\gamma\rho} + \varLambda^\rho_{\beta\gamma}F_{\alpha\rho} + \varLambda^\rho_{\gamma\alpha}F_{\beta\rho},$$

$$G^{\mu\alpha}_{\ ;\mu} \equiv \varLambda^\alpha_{\mu\nu}G^{\mu\nu}.$$

Ici l'indice de généralité du système est

$$2 + 48 - (n + 16) = 12;$$

il est *inférieur* à l'indice de généralité de l'ensemble des deux systèmes régissant, dans l'ancienne Physique, la gravitation et l'électromagnétisme.

II. *Point de vue du démiurge.* — Pour le démiurge, les relations (1) ne sont pas des *équations*, mais des *identités*, et il cherche à restreindre son schéma géométrique par l'introduction de n nouvelles équations de condition linéaires par rapport aux dérivées de la torsion et de caractère invariant.

Pour lui deux Univers sont équivalents si l'on peut passer de l'un à l'autre par une transformation ponctuelle telle qu'en deux points correspondants les *invariants différentiels* soient respectivement égaux. Ce qui caractérise un de ces Univers, c'est donc l'ensemble des relations qui existent entre les différents invariants différentiels. Parmi ces invariants, qui sont en nombre infini, il en existe un certain nombre, qu'on peut appeler fondamentaux, dont tous les autres peuvent se déduire par dérivation. Les n équations restrictives (2) doivent donc être interprétées comme établissant des conditions supplémentaires que doivent vérifier les relations qui existent entre les invariants différentiels fondamentaux.

On peut démontrer, presque sans calcul, à l'aide de la théorie des systèmes en involution, que *si les conditions d'involution trouvées par le physicien sont réalisées, le système différentiel qui donne les invariants différentiels fondamentaux en fonction de quatre d'entre eux est égale-*

---

1. On vérifie que ces deux relations ne contiennent aucune des dérivées par rapport à x^3 et x^4.

Here $r_1 = 0$ and $r_2 = 2$. Because of identities (1), the 2 relations between $\Lambda^\gamma_{\alpha\beta;1}$ and $\Lambda^\gamma_{\alpha\beta;2}$ will be

$$\Lambda^\mu_{12;\mu} = 0 \quad \text{et} \quad G^{34} - G^{43} - F^{34} = 0 \, {}^1.$$

The system is in involution, since one has the $12 = n - 12 + 2$ identities [3]

$$G^{\alpha\mu}_{\;;\mu} - F^{\alpha\mu}_{\;;\mu} \equiv -\Lambda^\nu_{\alpha\underline{\mu}}F_{\mu\nu},$$

$$F_{\alpha\beta;\gamma} + F_{\beta\gamma;\alpha} + F_{\gamma\alpha;\beta} \equiv \Lambda^\rho_{\alpha\beta}F_{\gamma\rho} + \Lambda^\rho_{\beta\gamma}F_{\alpha\rho} + \Lambda^\rho_{\gamma\alpha}F_{\beta\rho},$$

$$G^{\mu\alpha}_{\;;\mu} \equiv \Lambda^\alpha_{\mu\nu}G^{\mu\nu}.$$

Here, the generality index of the system is

$$2 + 48 - (n + 16) = 12;$$

it is *smaller* than the generality index of the set of two systems which determine gravitation and electromagnetism in the old Physics.

II. *The demiurge's point of view*

For the demiurge, relations (1) are not *equations* but *identities*, and he will try to restrict his geometrical scheme by introducing n new equations which will be linear with respect to the derivatives of the torsion and will have an invariant character.

For him, two Universes are equivalent if one can go from one to the other by a transformation such that at two corresponding points the *differential invariants* are equal. What characterizes one of the Universes is the set of relations that exist between the different differential invariants. Among these invariants, infinite in number, there are a certain number of them which one may call fundamental and from which all the others can be derived by differentiation. The n restrictive equations (2) must, therefore, be interpreted as establishing the supplementary conditions to be satisfied by the relations existing between the fundamental differential invariants.

It can be proven, almost without calculation, by using the theory of systems in involution, that *if the involution conditions found by the physicist are realized, the differential system giving the fundamental*

---

2. La remarque de Cartan au sujet de 12 identités liant les premiers membres des équations de champ va jouer un rôle important immédiatement après.

ment en involution et déterministe, et que son indice de généralité est $r_2 + 32 - n$, *comme pour le physicien.*

REMARQUE. — La démonstration précédente est indispensable. En effet nous avons supposé que le système différentiel du physicien était en involution et déterministe, *mais cela supposait nécessairement l'existence préalable de l'Univers.* Il est vrai qu'on peut regarder les équations (2) comme des équations aux dérivées partielles par rapport aux 16 fonctions inconnues $h_{s\beta}$; ce système est certainement en involution si les conditions énoncées plus haut sont réalisées, mais, *à ce point de vue*, il n'est pas déterministe et sa solution dépend de quatre fonctions arbitraires des quatre variables indépendantes; *cela ne nous donne par suite aucun moyen de savoir quel est le degré de généralité du schéma géométrique, considéré dans ce qu'il a d'essentiel*, c'est-à-dire d'indépendant du choix des variables. Ce n'est que l'étude du système qui peut nous donner ce degré de généralité. *Il se peut* du reste qu'on puisse établir un raisonnement a priori établissant les résultats précédents.

AUTRE REMARQUE. — Les 16 identités (1) qui, pour le physicien, expriment des lois de la nature, peuvent être interprétées géométriquement par le physicien qui sait être dans un espace riemannien avec l'existence d'un parallélisme *de proche en proche* (espace riemannien pouvant comporter *courbure* et *torsion*). Les lois (1) ne lui permettent pas à elles seules de reconnaître que son espace est sans courbure; elles expriment simplement que, *s'il y a une courbure*, les composantes R_{ijkh} de cette courbure satisfont aux mêmes relations de symétrie que les composantes du tenseur de Riemann dans un espace riemannien sans torsion. Les 22 équations d'Einstein et les 16 équations (1) pourraient donc être prises comme point de départ dans une théorie physique reposant sur la notion générale d'espace riemannien doué de courbure et de torsion, mais pour que la Physique soit déterministe, il faudrait ajouter d'autres équations faisant intervenir la courbure. On pourrait ainsi peut-être obtenir une théorie ayant un indice de généralité supérieur à 12 [3].

---

3. C'est là une remarque incidente qui attire l'attention sur un schéma avec courbure et torsion généralisant la Relativité générale d'Einstein sur lequel Cartan avait déjà insisté dans son grand mémoire des *Annales Ec. Normale* de 1923 [4].

Ce point de vue a été repris et développé par plusieurs auteurs au cours de ces dernières années.

differential invariants as functions of four of them is also deterministic and in involution and that its generality index is $r_2 + 32 - n$, *as in the case of the physicist.*

REMARK. — The preceding proof is essential. We have assumed that the physicist's differential system was deterministic and in involution, *but this necessarily assumes the previous existence of the Universe.* It is true that equations (2) can be regarded as partial differential equations with respect to the 16 unknown functions $h_{s\beta}$; this system is certainly in involution if the conditions stated above are realized, but *from this point of view* it is not deterministic and its solution depends on four arbitrary functions of the four independent variables. *It is then by no means possible to determine the degree of generality of the geometrical scheme, at its most fundamental level,* namely its independence of the choice of variables. Only the study of the system itself can give us this degree of generality. *It may be possible* to set up an " a priori " reasoning to prove the previous results.

ANOTHER REMARK. — The 16 identities (1) which, for the physicist, express laws of nature can be geometrically interpreted by a physicist who knows that he is in a Riemannian space with a " local " parallelism (Riemannian space with *curvature* and *torsion*). The laws (1) alone do not allow him to recognize that his space is curvature-free; they simply state that, *if curvature is present*, the components R_{ijkh} of this curvature satisfy the same symmetry relations as the components of the Riemann tensor in a torsion-free Riemannian space. Thus Einstein's 22 equations and the 16 equations (1) might be taken as a starting point for a physical theory which would rely on the general notion of Riemannian space endowed with curvature and torsion but in order that its physics be deterministic, other equations would have to be added to bring in the curvature. In this way, one could perhaps obtain a theory with a generality index larger than 12. [3]

Un article de synthèse à ce sujet a paru dans *Rev. of Modern Physics*, vol. 48, 1976, pp. 393-416, par F. W. Hehl, P. von der Heyde, G. D. Kerlick, sous le titre: General relativity with spin and torsion.

V. Indications sur la méthode permettant
de former le système invariant le plus général

Bornons-nous à considérer dans les premiers membres des équations à déterminer, l'ensemble des termes linéaires par rapport aux dérivées de la torsion. Nous supposerons, ce qui est permis, que tout est rapporté au 4-Bein.

Les 96 dérivées $\Lambda^k_{ij;h}$ sont liées par 16 relations linéaires (abstraction faite de termes quadratiques en $\Lambda^\gamma_{\alpha\beta}$). Il en reste donc 80. Ces 80 quantités constituent un tenseur *imprimitif*, décomposable en un certain nombre de tenseurs *primitifs*. Un tenseur primitif est un ensemble de p quantités qui subissent une substitution linéaire quand on effectue une rotation sur le 4-Bein mais tel qu'on ne puisse pas trouver un système de $q < p$ combinaisons linéaires indépendantes de ces quantités subissant, pour leur propre compte, une substitution linéaire par une rotation de 4-Bein. D'après un théorème général tout tenseur imprimitif est décomposable en tenseurs primitifs, de manière que chaque composante du tenseur imprimitif donné soit, d'une manière et d'une seule, la somme de composantes des tenseurs primitifs composants.

Les 80 dérivées $\Lambda^k_{ij;h}$ indépendantes se décomposent en

1° un scalaire $\Lambda^k_{\underline{kh};h}$;

2° deux tenseurs antisymétriques à 6 composantes

$$\text{a) } \Lambda^k_{ij;k}; \qquad \text{b) } \Lambda^j_{i\underline{k};k} - \Lambda^i_{j\underline{k};k};$$

3° trois tenseurs symétriques à 9 composantes

$$\text{a) } \phi_{i;j} + \phi_{j;i} - \frac{1}{2}\varepsilon_{ij}\phi_{\underline{k};k}; \quad \text{b) } \psi_{i;j} + \psi_{j;i}\,^{[1]};$$

$$\text{c) } \Lambda^j_{i\underline{k};k} + \Lambda^i_{j\underline{k};k} - \frac{1}{2}\varepsilon_{ij}\Lambda^h_{h\underline{k};k};$$

4° un tenseur à 10 composantes;

5° un tenseur à 30 composantes.

Les tenseurs à 6 composantes ne sont primitifs que dans le domaine réel; ils sont imprimitifs dans le domaine complexe (ils seraient impri-

---

[1]. On a posé $\psi_1 = S^{234}$ etc.; d'après les 16 identités de Bianchi, l'expression $\psi_{\underline{k};k}$ ne contient plus de dérivées premières.

V. Outline of a Method for Forming
the Most General Invariant System

On the left hand sides of the equations to be determined, let us only consider just the set of terms linear with respect to the derivatives of the torsion. We shall assume that everything is referred to a 4-Bein.

The 96 derivatives $\Lambda^k_{ij;h}$ are related by 16 linear relations (apart from terms quadratic in $\Lambda^\gamma_{\alpha\beta}$). Thus, 80 of them still remain. These 80 quantities form an *non-primitive* tensor, decomposable into a certain number of *primitive* tensors. A primitive tensor is a set of p quantities which undergo a linear substitution when a rotation of the 4-Bein is performed, such that no system of $q < p$ linearly independent combinations of these quantities can be found which undergo a linear substitution of their own. According to a general theorem, any non-primitive tensor can be decomposed into primitive tensors in such a way that each component of the given non-primitive tensor is, in one and only one way, the sum of the components of the primitive tensors.

The 80 independent derivatives $\Lambda^k_{ij;h}$ decompose into

1. a scalar $\Lambda^k_{kh;h}$;
2. two skew-symmetric tensors with 6 components

$$\text{a) } \Lambda^k_{ij;k}; \qquad \text{b) } \Lambda^j_{i\underline{k};k} - \Lambda^i_{j\underline{k};k};$$

3. three symmetric tensors with 9 components

$$\text{a) } \phi_{i;j} + \phi_{j;i} - \frac{1}{2}\varepsilon_{ij}\phi_{k;k}; \quad \text{b) } \psi_{i;j} + \psi_{j;i}{}^{[1]};$$

$$\text{c) } \Lambda^j_{i\underline{k};k} + \Lambda^i_{j\underline{k};k} - \frac{1}{2}\varepsilon_{ij}\Lambda^h_{h\underline{k};k};$$

4. a tensor with 10 components;
5. a tensor with 30 components.

The tensors with 6 components are primitive only in the real domain: they are non-primitive in the complex domain (they would be non-primitive in the real domain if the fundamental quadratic form

---

1. Here $\psi_1 = s^{234}$ etc.; Because of the Bianchi identities, $\psi_{\underline{k};k}$ no longer contains any first derivatives.

mitifs dans le domaine réel si la forme quadratique fondamentale était définie). Cela est la source de complications fâcheuses. En effet le tenseur antisymétrique à 6 composantes le plus général τ_{ij} dépend de quatre constantes arbitraires, à savoir

$$\tau_{12} = a\Lambda^k_{12;k} + b\Lambda^k_{34;k} + c(\Lambda^2_{1k;k} - \Lambda^1_{2k;k}) + d(\Lambda^4_{3k;k} - \Lambda^3_{4k;k}).$$

Au contraire le tenseur symétrique (à 9 composantes, c'est-à-dire admettant un tenseur contracté nul) le plus général est

$$\alpha\left[\phi_{i;j} + \phi_{j;i} - \frac{1}{2}\varepsilon_{ij}\phi_{\underline{k};k}\right] + \beta[\psi_{i;j} + \psi_{j;i}]$$
$$+ \gamma\left[\Lambda^j_{i\underline{k};k} + \Lambda^i_{j\underline{k};k} - \frac{1}{2}\varepsilon_{ij}\Lambda^h_{h\underline{k};k}\right].$$

Cela posé, tout système d'équations invariantes s'obtiendra nécessairement en annulant un ou plusieurs des tenseurs primitifs trouvés. En fait un examen plus approfondi montre qu'on ne doit annuler ni le tenseur à 10 composantes, ni le tenseur à 30 composantes, *du moins si on veut arriver à un indice de généralité au moins égal à* 12.

L'indice de généralité maximum est 18, correspondant à 15 = 6 + 9 équations; puis vient l'indice de généralité 16 correspondant à 16 = 1 + 6 + 9 équations (mais la possibilité d'un système en involution de cette nature n'est pas démontrée); enfin vient l'indice de généralité 12 avec 22 = 1 + 6 + 6 + 9 équations.

Pour achever la position mathématique du problème, il est à peu près nécessaire de faire une hypothèse sur la forme des parties des équations indépendantes des dérivées $\Lambda^\gamma_{\alpha\beta;\mu}$.

HYPOTHÈSE III. — *Les équations à déterminer sont quadratiques par rapport aux composantes de la torsion.*

Cette dernière hypothèse pose le problème de la recherche des formes quadratiques ayant la même structure tensorielle que les tenseurs dont il a été question plus haut. À cet égard il existe

1° 5 scalaires indépendants, à savoir

$$\phi^2_k; \ \phi_k\psi_k; \ \psi^2_k; \ (\Lambda^i_{jk})^2; \ \sum_i(\Lambda^i_{23}\Lambda^i_{14} + \Lambda^i_{31}\Lambda^i_{24} + \Lambda^i_{12}\Lambda^i_{34});$$

2° 3 tenseurs antisymétriques à 6 composantes, à savoir

$$\phi_i\psi_j - \phi_j\psi_i; \ \Lambda^k_{ij}\phi_k; \ \Lambda^k_{ij}\psi_k;$$

3° 10 tenseurs symétriques à 9 composantes, qu'il est inutile d'écrire.

were definite). This is the source of annoying complications. Indeed the most general skew-symmetric tensor τ_{ij} with 6 components depends on four arbitrary constants, namely

$$\tau_{12} = a\Lambda^k_{12;k} + b\Lambda^k_{34;k} + c(\Lambda^2_{1\underline{k};k} - \Lambda^1_{2\underline{k};k}) + d(\Lambda^4_{3\underline{k};k} - \Lambda^3_{4\underline{k};k}).$$

On the other hand, the most general symmetric tensor (with 9 components, i.e. having its contraction equal to zero) is

$$\alpha\left[\phi_{i;j} + \phi_{j;i} - \frac{1}{2}\varepsilon_{ij}\phi_{k;k}\right] + \beta[\psi_{i;j} + \psi_{j;i}]$$
$$+ \gamma\left[\Lambda^j_{i\underline{k};k} + \Lambda^i_{j\underline{k};k} - \frac{1}{2}\varepsilon_{ij}\Lambda^h_{h\underline{k};k}\right].$$

This being granted, any system of invariant equations must necessarily be obtained by setting equal to zero one or several of the primitive tensors. In fact, a more careful examination shows that neither the 10 component tensor nor the 30 component tensor can be zero, *at least if one wants to get a generality index at least equal to* 12.

The highest generality index is 18, corresponding to $15 = 6 + 9$ equations. Then comes an index of 16, corresponding to $16 = 1 + 6 + 9$ equations (though existence of a system in involution of that nature has not been proven). Finally one has an index of 12, with $22 = 1 + 6 + 6 + 9$ equations.

To complete the mathematical situation, it is almost necessary to make an assumption about the form of those parts of the equations which do not depend on the derivatives $\Lambda^\gamma_{\alpha\beta;\mu}$.

ASSUMPTION III. — *The equations to be determined are quadratic with respect to the components of the torsion.*

This last assumption poses the problem of the determination of those quadratic forms which have the same tensorial structure as the tensors discussed above. In this respect, there exist

1. 5 independent scalars, namely

$$\phi^2_k; \quad \phi_k\psi_k; \quad \psi^2_k; \quad (\Lambda^i_{jk})^2; \quad \sum_i (\Lambda^i_{23}\Lambda^i_{14} + \Lambda^i_{31}\Lambda^i_{24} + \Lambda^i_{12}\Lambda^i_{34});$$

2. 3 skew-symmetric tensors with 6 components, namely

$$\phi_i\psi_j - \phi_j\psi_i; \quad \Lambda^k_{ij}\phi_k; \quad \Lambda^k_{ij}\psi_k;$$

3. 10 symmetric tensors with 9 components which are not worth writing down.

VIII

8-XII-29

Verehrter Herr Cartan!

Ich bin gerührt und erfreut darüber, dass Sie sich schon so viele Mühe mit dem Problem gegeben haben. Ich beantworte einstweilen nur den Brief, die Antwort auf die Abhandlung folgt später.

Sie stellen die Gleichungen auf

$$R_{ik} = 0, \tag{1}$$

$$\Lambda^{\mu}_{\alpha\beta;\mu} = \phi_{\alpha,\beta} - \phi_{\beta,\alpha} = C\, h\, S^{\gamma\delta\mu}\phi_{\mu}, \tag{2}$$

$$(h\, S^{\alpha\beta\mu})_{,\mu} = C'\, h\, S^{\alpha\beta\mu}\phi_{\mu}. \tag{3}$$

Die Kompatibilität dieser Gleichungen soll durch den integrierenden Ansatz

$$\phi_{\alpha} = \frac{1}{C'}\, \frac{1}{\psi}\, \frac{\partial\psi}{\partial x^{\alpha}}, \tag{4}$$

$$h\, S^{\alpha\beta\gamma} = \frac{1}{\psi}\, \frac{\partial\chi}{\partial x^{\delta}}, \tag{5}$$

gezeigt werden.

(4) hat aber das Verschwinden von $\phi_{\alpha,\beta} - \phi_{\beta,\alpha}$ zur Folge, also gemäss (2) das Verschwinden von S···. Dann sagt (3) nichts mehr aus. Der Ansatz (4) ist also nicht möglich und die Kompatibilität von (1), (2), (3) nicht erwiesen.

Eine allgemeine Bemerkung: Wenn Gleichungen allgemein integrierbar sind wie z.B. $\Lambda^{\alpha}_{\mu\nu;\alpha} = 0 \left(\text{durch } \phi_{\mu} = \dfrac{\partial\log\psi}{\partial x^{\mu}} \right)$, so soll man die *integrierten* Gleichungen bei der Abzählung heranziehen. Es scheint mir richtig zu sagen, dass dies nicht 6 sondern 4 Gleichungen sind.

A VIII

8.XII.29

Dear M. Cartan,

I am both touched and delighted that you have taken so many pains over the problem. For the present, I am responding only to your letter; an answer to your Note will follow later.

You set up the equations

$$R_{ik} = 0, \tag{1}$$

$$\Lambda^{\mu}_{\alpha\beta;\mu} = \phi_{\alpha,\beta} - \phi_{\beta,\alpha} = C\, h\, S^{\gamma\delta\mu}\phi_{\mu}, \tag{2}$$

$$(h\, S^{\alpha\beta\mu})_{,\mu} = C'\, h\, S^{\alpha\beta\mu}\phi_{\mu}. \tag{3}$$

The compatibility of these equations is to be shown by the integrating assumption

$$\phi_{\alpha} = \frac{1}{C'}\frac{1}{\psi}\frac{\partial\psi}{\partial x^{\alpha}}, \tag{4}$$

$$h\, S^{\alpha\beta\gamma} = \frac{1}{\psi}\frac{\partial\chi}{\partial x^{\delta}}, \tag{5}$$

But from (4) it follows that $\phi_{\alpha,\beta} - \phi_{\beta,\alpha}$ vanishes and thus from (2) $S^{\cdots}$ vanishes. So (3) no longer has any content. Assumption (4) is thus not allowable and the compatibility of (1), (2), and (3) remains unproven.

A general remark: When equations are generally integrable, as for example $\Lambda^{\alpha}_{\mu\nu;\alpha} = 0 \left(\text{with } \phi_{\mu} = \dfrac{\partial\log\psi}{\partial x^{\mu}}\right)$, then the *integrated* equations should be included in the count. It seems correct to me to say that these are not 6 but rather 4 equations.

Man könnte Ihren Fall verallgemeinernd folgende Feldgleichungen schreiben

$$R_{ik} = 0,$$

$$\phi_\alpha = A(\psi,\chi)\frac{\partial \psi}{\partial x^\alpha} + B(\psi,\chi)\frac{\partial \chi}{\partial x^\alpha},$$

$$h\, S^{\sigma\tau\rho}\, \delta_{\sigma\tau\rho\alpha} = C(\psi, \chi)\frac{\partial \psi}{\partial x^\alpha} + D(\psi,\chi)\frac{\partial \chi}{\partial x^\alpha},$$

wobei A, B, C, D gegebene Funktionen zweier Argumente sind. Man hat dann 10 + 4 + 4 Gleichungen mit 4 Identitäten für 16 + 2 Variable, was gestattet ist.

Ich denke aber, dass dies eine zu schwache Über-Bestimmung ist. Bei meinem System hat man 7 Identitäten statt nur 4. Darin sehe ich das Wesentliche.

Wenn man sich mit 4 Identitäten begnügen will kann man sich des Hamilton' schen Prinzipes bedienen

$$\delta\left\{\int \mathscr{H}\, h\, d\tau\right\} = 0$$

wobei $\mathscr{H}$ ein aus den $\varLambda$ quadratisch gebildeter Skalar ist. Für $\mathscr{H}$ gibt es hiebei sehr viele Möglichkeiten. Ich bin aber der Überzeugung dass man eine stärkere Determinierung verlangen muss.

Auch Ihr System

$$R_{ik} = 0 \quad a\phi_1 + bS_1 = \frac{\partial \psi}{\partial x^1}, \quad \varLambda^\beta_{\alpha\mu;\mu} + \ldots = A^{\alpha\beta} - \frac{1}{4}A^{\mu\mu}(g^{\alpha\beta})$$
$$\ldots\ldots\ldots\ldots\ldots\ldots$$
$$\ldots\ldots\ldots\ldots\ldots\ldots$$

scheint mir nicht acceptabel. Es sind 10 + 4 + 9 Gleichungen mit 4 Identitäten für 16 + 1 Feldvariable. Also sind es zwei Gleichungen zu viel, solange nicht die Existenz zweier weiterer Identitäten aufgezeigt ist.

Endlich das System von 9 + 1 + 6 Gleichungen, dessen skalare Gleichung Sie $\phi_{\mu;\mu} = C$ geschrieben haben. Hier ist überhaupt keine Identität ersichtlich, sodass ich die Kompatibilität nicht einsehe. Aber selbst wenn 4 Identitäten ausfindig gemacht würden, so wäre die Determination schwächer als bei meinem System. *Und der Determinations-Grad ist gerade das heuristische Prinzip*; nach meiner

Generalising your case, one can write the following field equations:

$$R_{ik} = 0,$$

$$\phi_\alpha = A(\psi,\chi) \frac{\partial \psi}{\partial x^\alpha} + B(\psi,\chi) \frac{\partial \chi}{\partial x^\alpha},$$

$$h\, S^{\sigma\tau\rho}\, \delta_{\sigma\tau\rho\alpha} = C(\psi,\chi) \frac{\partial \psi}{\partial x^\alpha} + D(\psi,\chi) \frac{\partial \chi}{\partial x^\alpha},$$

where A, B, C, D are given functions of two variables. Then there are 10 + 4 + 4 equations with 4 identities for 16 + 2 variables, which is allowed. But I think this is too weak an overdetermination. In my system there are 7 identities instead of only 4. This is where I see the essential point.

If one is willing to be satisfied with 4 identities then one can use Hamilton's Principle

$$\delta \left\{ \int \mathscr{H}\, h\, d\tau \right\} = 0$$

where $\mathscr{H}$ is a scalar quadratic in Λ. There are very many possibilities for $\mathscr{H}$, but I am convinced that one must demand a stronger determination.

Your system:

$$R_{ik} = 0 \qquad a\phi_1 + bS_1 = \frac{\partial \psi}{\partial x^1}, \qquad \Lambda^\beta_{\underline{\alpha\mu};\mu} + \ldots = A^{\alpha\beta} - \frac{1}{4} A^{\mu\mu}(g^{\alpha\beta})$$

$$\ldots\ldots\ldots\ldots\ldots$$
$$\ldots\ldots\ldots\ldots\ldots$$

also appears unacceptable to me. There are 10 + 4 + 9 equations with 4 identities for 16 + 1 field variables. Thus there are two equations too many, unless one can show the existence of two more identities.

Finally, the system of 9 + 1 + 6 equations, whose scalar equation you have written as $\phi_{\mu;\mu} = C$. Here no identity is obvious at all, so I cannot see the compatibility. But even if 4 identities were to be discovered, the degree of determination would be weaker than in my system. *And the degree of determination is precisely the heuristic principle*; in my opinion, a theory is the more valuable the more strongly

Ansicht ist eine Theorie desto wertvoller, je stärker sie die Möglichkeiten einschränkt, ohne mit der Wirklichkeit in Konflikt zu kommen. Es ist wie bei einem Steckbrief, welcher einen Verbrecher charackterisieren soll; je *mehr Zutreffendes* er aussagt, desto mehr ist er wert. Der Determinationsgrad scheint mir durch die Zahl der (voneinander unabhängigen) Identitäten ausgedrückt zu sein (wohin vorausgestzt ist, dass diese gleich N — 12 ist).

Sie stellen die Frage auf, wie der Fall zu interpretieren sei, das die Zahl der unabhängigen Identitäten grösser sei als N — 12, wenn N die Zahl der Gleichungen ist. Ich habe nicht verstehen können, wie Sie es meinen, wenn Sie sagen, dass im Falle meiner ursprünglichen Gleichung ($\Lambda^{\alpha}_{\mu\nu;\alpha} = 0$) die Zahl der Identitäten nur zwei grösser sei als die Zahl N — 12. Ich möchte dazu bemerken, dass von den Identitäten

$$\frac{\partial F_{\mu\nu}}{\partial x^{\sigma}} + \frac{\partial F_{\nu\sigma}}{\partial x^{\mu}} + \frac{\partial F_{\sigma\mu}}{\partial x^{\nu}} \equiv 0$$

und

$$\frac{\partial[\psi h\, K^{\mu\nu}]}{\partial x^{\nu}} \equiv 0$$

(wobei $K^{\mu\nu}$ die Abkürzung für einen Ausdruck ist, in dem neben $G^{\mu\nu}$ und $F^{\mu\nu}$ auch F_{μ} auftritt) nur 6 zählen und nicht alle 8. Nennt man nämlich $F_{\mu\nu\sigma}$ die linke Seite des ersten Systems K^{μ} wie linke Seite des zweiten, so verschwinden $(F_{\mu\nu\sigma}\delta^{\mu\nu\sigma})_{,\tau}$ und $K^{\mu}_{,\mu}$ identisch, d.h. unabhängig davon wie die Grössen $F_{\mu\nu\sigma}$ und $K^{\mu\nu}$ sich durch die $F^{\mu\nu}$ und $G^{\mu\nu}$ ausdrücken. Die zweite Identität ist aber (auch in der Form $(hK^{\mu\nu})_{,\nu} + hK^{\mu\nu}\phi_{\nu} \equiv 0$) keine reine identische Beziehung zwischen den linken Seiten der ursprünglichen Feldgleichungen, weil darin die F_{μ} auftreten. Ich weiss daher nicht, wie man sich ohne Einführung von ψ von der Kompatibilität der Gleichungen überzeugen soll. Was der Fall N — 12 < Zahl der Identitäten bedeutet, ist mir nicht völlig deutlich geworden. Ich vermute aber, dass dann der Determinismus unvollständig ist.

Alles Weitere nach Studium Ihrer genaueren Darlegungen. Herzlich grüsst Sie Ihr

A. Einstein

it restricts possibilities, without coming into conflict with reality. It is like a wanted poster which is supposed to characterize a criminal; the *more precisely* it points him out the better. It seems to me that the degree of determination is to be expressed by the number of (independent) identities (where it is assumed that this is equal to N − 12).

You propose the question of the interpretation of the case when the number of independent identities is larger than N − 12, where N is the number of equations. I have not been able to understand what you mean when you say that, in the case of my original equations ($\Lambda^{\alpha}_{\mu\nu;\alpha} = 0$), the number of identities is only two greater than the number N − 12. I would remark that, with respect to the identities,

$$\frac{\partial F_{\mu\nu}}{\partial x^{\sigma}} + \frac{\partial F_{\nu\sigma}}{\partial x^{\mu}} + \frac{\partial F_{\sigma\mu}}{\partial x^{\nu}} \equiv 0$$

and

$$\frac{\partial \left[\psi h\, K^{\mu\nu}\right]}{\partial x^{\nu}} \equiv 0$$

(where $K^{\mu\nu}$ is an abbreviation for an expression in terms of $G^{\mu\nu}$ $F^{\mu\nu}$, and F_{μ}) only 6 count and not all 8. For taking $F_{\mu\nu\sigma}$ for the left hand side of the first system and K^{μ} for the left side of the second, then $(F_{\mu\nu\sigma}\delta^{\mu\nu\sigma\tau})_{,\tau}$ and $K^{\mu}_{,\mu}$ vanish identically, i.e., independently of how the quantities $F_{\mu\nu\sigma}$ and K^{μ} are expressed in terms of $F^{\mu\nu}$ and $G^{\mu\nu}$. However, the second identity (even in the form $(hK^{\mu\nu})_{,\nu} + hK^{\mu\nu}\phi_{\nu} \equiv 0$) is not a pure identical relation between the left hand sides of the original field equations since the F_{μ} enter in. But I don't know how one could guarantee the compatibility of the equations without introducing the ψ. What the case N − 12 < the number of identities means is not entirely clear to me. I suppose that in this case the system is underdetermined.

More after a study of your detailed exposition. Kind regards.

Yours,

A. Einstein

P.S. Ich habe keine klare Vorstellung davon, welches die Mannigfaltigkeit der Lösungen in solchem Falle ist, dass weniger als N − 12 Identitäten vorhanden sind (z.B. $\Lambda^{\alpha}_{\underline{\mu}\nu;\nu} = 0$). Eine deterministische Form gibt es dann nicht.

Ich studiere eifrig an Ihrem Manuskript *und bin sehr dankbar für Ihr Interesse an dem Problem.*

P.S. I have no clear idea which of the solution manifolds exists in the case that there are less than $N - 12$ identities (e.g. $\Lambda^\alpha_{\underline{\mu\nu};\nu} = 0$). There is no deterministic form in this case.

I am eagerly studying your manuscript *and am very thankful for your interest in the problem.*

IX

Le Chesnay,
le 13 décembre 1929

Cher et illustre Maître,

J'ai bien reçu votre lettre et je l'ai lue et méditée. Ce que vous me dites au sujet du degré d'indétermination m'a intéressé, car j'avais pensé, pour des raisons que j'expose dans ma note, mais qui sont un peu des raisons de sentiment, que votre système de 22 équations était peut-être *trop* déterminé et que sa solution n'avait pas un degré de généralité suffisant; mais vous êtes à cet égard infiniment plus compétent que moi. Je crois néanmoins que ma notion d'indice de généralité, qui a un sens tout à fait précis, peut rendre des services. Si votre système par exemple, ne dépendait, au sens où je l'indique dans ma note, que de 3 ou 4 fonctions arbitraires de 3 variables, il ne serait certainement pas assez général. Mon indice de généralité varie évidemment en sens inverse du nombre des identités, du moins dans les cas où ce nombre est $N - 12$.

Vous m'indiquez, relativement à certains des systèmes que je vous avais signalés, des objections qui ne sont dues, j'en suis persuadé, qu'à la mauvaise rédaction de ma lettre. Ainsi pour le système de 22 équations, que je peux écrire

$$R_{ij} = 0, \tag{1}$$

$$F_{\alpha\beta} \equiv \phi_{\alpha,\beta} - \phi_{\beta,\alpha} - C(\phi_\alpha S_\beta - \phi_\beta S_\alpha) = 0, \tag{2}$$

$$G_{\alpha\beta} \equiv S_{\alpha,\beta} - S_{\beta,\alpha} - C'(\phi_\alpha S_\beta - \phi_\beta S_\alpha) = 0, \tag{3}$$

je n'ai jamais eu l'idée de démontrer la compatibilité en intégrant les équations (2) et (3), surtout en les intégrant d'une manière fausse. La compatibilité résulte pour moi de l'existence de $n - 12 + r_2 = 22 - 12 + 2 = 12$ identités, qui sont, en dehors des 4 identités de Bianchi, les 8 suivantes

$$\frac{\partial F_{\beta\gamma}}{\partial x_\alpha} + \frac{\partial F_{\gamma\alpha}}{\partial x_\beta} + \frac{\partial F_{\alpha\beta}}{\partial x_\gamma} - C(S_\alpha F_{\beta\gamma} + \dots - \phi_\alpha G_{\beta\gamma} \dots) \equiv 0,$$

$$\frac{\partial G_{\beta\gamma}}{\partial x_\alpha} + \frac{\partial G_{\gamma\alpha}}{\partial x_\beta} + \frac{\partial G_{\alpha\beta}}{\partial x_\gamma} - C'(S_\alpha F_{\beta\gamma} + \dots - \phi_\alpha G_{\beta\gamma} \dots) \equiv 0.$$

A IX

Le Chesnay,
13 December 1929

Cher et illustre Maître,

I have received your letter and I have read and thought about it. What you say about the degree of indetermination interests me because I thought, for reasons that I outline in my note but which are a bit personal, that your system of 22 equations was perhaps *too* determined and that its solution did not have a sufficient degree of generality. But in this respect, you are infinitely more competent than I. Nevertheless, I think, that my concept of generality index, which has a very precise meaning, may be of some help. If your system, for instance, depended, in the sense I indicate in my note, only on 3 or 4 arbitrary functions of 3 variables, it would certainly not be general enough. My generality index varies, of course, in inverse proportion to the number of identities, at least when this number is $N - 12$.

You raise objections to some of the systems I mention, objections that I am sure, are due only to the bad phrasing of my letter. Thus, in the case of the system of 22 equations, which I write

$$R_{ik} = 0, \qquad (1)$$

$$F_{\alpha\beta} \equiv \phi_{\alpha,\beta} - \phi_{\beta,\alpha} - C(\phi_\alpha S_\beta - \phi_\beta S_\alpha) = 0, \qquad (2)$$

$$G_{\alpha\beta} \equiv S_{\alpha,\beta} - S_{\beta.\alpha} - C'(\phi_\alpha S_\beta - \phi_\beta S_\alpha) = 0, \qquad (3)$$

I never intended to prove their compatibility by integrating equations (2) and (3), and especially not by integrating them incorrectly. Their compatibility results rather, for me, from the existence of $n - 12 + r_2 = 22 - 12 + 2 = 12$ identities which apart from the 4 Bianchi identities, are the following 8:

$$\frac{\partial F_{\beta\gamma}}{\partial x_\alpha} + \frac{\partial F_{\gamma\alpha}}{\partial x_\beta} + \frac{\partial F_{\alpha\beta}}{\partial x_\gamma} - C(S_\alpha F_{\beta\gamma} + \ldots - \phi_\alpha G_{\beta\gamma} \ldots) \equiv 0,$$

$$\frac{\partial G_{\beta\gamma}}{\partial x_\alpha} + \frac{\partial G_{\gamma\alpha}}{\partial x_\beta} + \frac{\partial G_{\alpha\beta}}{\partial x_\gamma} - C'(S_\alpha F_{\beta\gamma} + \ldots - \phi_\alpha G_{\beta\gamma} \ldots) \equiv 0.$$

Je crois bien vous avoir fait remarquer que, dans votre système de 22 équations, figurent les 12 équations (2) et (3) avec la valeur 0 de la constante C et la valeur − 1 de la constante C′, et avoir ajouté que, *dans le cas particulier* C = 0, on peut intégrer non seulement les équations (2), mais encore les équations (3), par l'introduction de deux nouvelles fonctions ψ et χ. L'intégration peut du reste se faire aussi si C ≠ 0; il n'y a qu'à poser

$$\Sigma_\alpha = S_\alpha - \frac{C'}{C}\,\phi_\alpha$$

et à prendre ϕ_α et Σ_α comme fonctions inconnues; les équations (2) et (3) conservent la même forme, mais la nouvelle constante C′ est nulle.

J'aurais bien à dire quelque chose sur la généralisation que vous me proposez avec des fonctions arbitraires de ψ et χ, mais cela n'en vaut pas la peine.

J'arrive maintenant au système de 15 équations dont je vous avais parlé; il se compose

1° des 6 équations

$$F_{\alpha\beta} \equiv \frac{\partial(aS_\alpha + b\phi_\alpha)}{\partial x_\beta} - \frac{\partial(aS_\beta + b\phi_\beta)}{\partial x_\alpha} = 0$$

2° de 9 équations dont les premiers membres sont assujettis à la seule condition qu'ils forment un tenseur symétrique, avec un tenseur scalaire contracté nul.

Je n'ai jamais songé à faire entrer dans ce système les équations $R_{ik} = 0$, qui rendraient le système incompatible. Ce système de 15 équations est très intéressant au point de vue mathématique (il est entendu qu'il n'est pas assez déterminé pour la physique). Écrivons l'une des 9 dernières équations, par exemple

$$G_{12} = \Lambda^2_{1\mu;\mu} + \Lambda^1_{2\mu;\mu} + c(\phi_{1,2} + \phi_{2,1}) + d(S_{1,2} + S_{2,1}) - A_{12} = 0,$$

A_{12} ne dépendant que des $\Lambda^\gamma_{\alpha\beta}$ et non de leurs dérivées.

Dans le cas général, c'est-à-dire si l'on a

$$a - bd \neq 0, \qquad c \neq -\frac{1}{3},$$

I think I pointed out to you that, in your system of 22 equations, the 12 equations (2) and (3) are present, with the value 0 for the constant C and the value -1 for the constant C', and also that, *in this special case* C $= 0$, one can integrate not only equations (2), but also equations (3) by introducing the two new functions ψ and χ. The integration can be performed even if C $\neq 0$. One need only set

$$\Sigma_\alpha = S_\alpha - \frac{C'}{C}\,\phi_\alpha$$

and take ϕ_α and Σ_α as unknown functions. Then equations (2) and (3) keep the same form but the new constant C' is zero.

I have a few things about the generalization you propose to me, with arbitrary functions of ψ and χ but it is not worthwhile writing them down.

Now I come to the system of 15 equations that I mentioned to you; it is composed of

1) the 6 equations

$$F_{\alpha\beta} \equiv \frac{\partial(aS_\alpha + b\phi_\alpha)}{\partial x_\beta} - \frac{\partial(aS_\beta + b\phi_\beta)}{\partial x_\alpha} = 0$$

2) 9 equations the lefthand sides of which are subject to the single condition that they form a trace-free symmetric tensor.

I never intended to have the equations $R_{ik} = 0$ *enter into the system*, something which would make the system incompatible. This system of 15 equations is very interesting from the mathematical point of view (it is understood that it is not determined enough for physics). Let us write down one of the 9 last equations, for instance,

$$G_{12} = \Lambda^2_{1\mu;\mu} + \Lambda^1_{2\mu;\mu} + c(\phi_{1,2} + \phi_{2,1}) + d(S_{1,2} + S_{2,1}) - A_{12} = 0,$$

with A_{12} depending only on the $\Lambda^\gamma_{\alpha\beta}$, and not on their derivatives.
In the general case, that is, if

$$a - bd \neq 0, \qquad c \neq -\frac{1}{3},$$

the system is deterministic and in involution. But if $a - bd$ or $c + \frac{1}{3}$ is zero, the system is no longer deterministic nor in involution: one can

le système est déterministe et en involution. Mais si $a - bd$ ou $c + \dfrac{1}{3}$ est nul, le système cesse d'être déterministe et cesse aussi d'être en involution: on ne peut plus affirmer qu'il soit compatible, mais il n'est pas sûr non plus qu'il n'admette aucune solution.

Comme il me semble que nous ne sommes peut-être pas absolument d'accord sur le sens du mot *identités*, je me permets de vous dire d'une manière précise quel est le point de vue auquel me conduit la théorie des systèmes en involution.

Prenons N équations aux dérivées partielles linéaires du 1er ordre à p fonctions inconnues

$$F_\alpha = 0 \ (\alpha = 1, 2, ..., N)$$

et supposons que les premiers membres soient des fonctions linéaires indépendantes des dérivées des p fonctions inconnues par rapport à l'une quelconque des variables. Il existe alors un nombre *maximum*

$$N - p + r_1 + r_2$$

de combinaisons linéaires indépendantes des $F_{\alpha,\beta}$ jouissant de la propriété que dans chacune de ces combinaisons, les dérivées du second ordre s'éliminent d'elles-mêmes.

Cela posé, trois cas sont possibles:

1. Ce nombre maximum n'est pas atteint;

2. Ce nombre maximum est atteint, mais les $N - p + r_1 + r_2$ combinaisons trouvées (qui ne dépendent par conséquent que des fonctions inconnues et de leurs dérivées premières) ne sont pas toutes nulles si l'on tient compte des équations données $F_\alpha = 0$;

3. Le nombre maximum est atteint, mais les $N - p + r_1 + r_2$ combinaisons trouvées sont toutes nulles en tenant compte des équations données.

Dans ce troisième cas, je dis que les premiers membres des équations sont liés par $N - p + r_1 + r_2$ identités: c'est la condition nécessaire et suffisante pour que le système soit complètement intégrable.

Dans le 2e cas, le système est incompatible, ou du moins toutes les solutions du système satisfont nécessairement à de nouvelles équations aux dérivées partielles du premier ordre qui ne sont pas des conséquences *algébriques* des équations données. Comme c'est une éventualité que nous excluons, nous pouvons dire que le système est incompatible.

no longer affirm that it is compatible but then neither is it certain that it permits no solution.

As it seems we do not completely agree about the meaning of the word *identities*, allow me to tell you precisely the point of view I adopt by the theory of systems in involution.

Let us take N linear partial differential equations of first order in p unknown functions,

$$F_\alpha = 0 \ (\alpha = 1, 2, ..., N)$$

and let us assume that the left hand sides are linear functions of the derivatives of the p unknown functions and independent of the derivatives with respect to any one of the variables. Then there exists a maximum number $N - p + r_1 + r_2$ of independent linear combinations of the $F_{\alpha,\beta}$ having the property that, in any one of the combinations, the second order derivatives cancel out.

Three cases are then possible:

1. This maximum number is not reached;
2. This maximum number is reached, but the $N - p + r_1 + r_2$ combinations (which thus, depend only on the unknown functions and their first derivatives) are not all zero if one takes into account the given equation, $F_\alpha = 0$;
3. The maximum number is reached, but the $N - p + r_1 + r_2$ combinations are all zero because of the given equation.

In this third case I say that the left-hand sides of the equations are related by $N - p + r_1 + r_2$ identities: this is the necessary and sufficient condition for the system to be completely integrable.

In the second case, the system is incompatible, or at least all the solutions of the system necessarily satisfy new first order partial differential equations which are not algebraic consequences of the given equation. As this is a possibility that we exclude, we can say that the system is incompatible.

In the first case the system is not completely integrable, but one cannot assert without further examination that it is incompatible. One is obliged to consider the system obtained by adding to the given equation, the derived equations (the first order derivatives of the unknown functions being the new unknown functions) and then to see if the

Dans le 1er cas, le système n'est pas complètement intégrable, mais on ne peut affirmer sans plus ample examen qu'il soit incompatible. On est obligé de considérer le système obtenu en adjoignant aux équations données les équations dérivées (les dérivées du premier ordre des fonctions inconnues étant de nouvelles fonctions inconnues), puis de chercher si les conditions d'involution de ce nouveau système sont réalisées, et ainsi de suite. On est sûr de finir par tomber sur un des cas 3° ou 2°.

Le système de 16 équations dont je vous avais parlé, et dont je n'écris que quelques-unes,

$$X_{12} \equiv S_{1,2} - S_{2,1} + a(\phi_{1,2} - \phi_{2,1}) + b(\phi_{3,4} - \phi_{4,3}) - A_{12} = 0,$$

$$Z \equiv \phi_{\underline{\alpha},\alpha} - C = 0,$$

$$Y^{12} \equiv \Lambda^2_{1\underline{\alpha};\alpha} + \Lambda^1_{2\underline{\alpha};\alpha} + c(\phi_{\underline{1,2}} + \phi_{\underline{2,1}}) + d(S_{\underline{1,2}} + S_{\underline{2,1}}) - g^{12}[\Lambda^\mu_{\mu\underline{\alpha};\alpha} + \ldots] = 0,$$

rentre dans le second cas; il y a 4 combinaisons linéaires des dérivées des premiers membres, d'où les dérivées du second ordre s'éliminent; l'une de ces combinaisons est

$$Y_{1\underline{\alpha};\alpha} - \frac{3c+1}{2}Z_{;1} - dX_{1\underline{\alpha};\alpha} - \frac{1+c-ad}{6}h\,(X_{\underline{23};4} + X_{\underline{34};2} + X_{\underline{42};3}).$$

Si les quantités A_{12}, C, etc., sont quelconques, le système est impossible, mais je ne sais pas si on peut les choisir de manière que le système soit complètement intégrable.

Ce que vous me dites au sujet du nombre d'identités *réelles* de votre système se ramène peut-être à une question de convention. Quand je dis qu'il y a 12 identités, je sous-entends que je ne fais intervenir que les équations telles qu'elles sont, sans les modifier par l'introduction d'une fonction auxiliaire. Mes 12 identités, à savoir

$$G^{\alpha\mu}_{;\mu} + F^{\alpha\mu}_{;\mu} + \Lambda^\rho_{\underline{\alpha\mu}}F_{\mu\rho} \equiv 0,$$

$$G^{\mu\alpha}_{;\mu} + \Lambda^\alpha_{\rho\sigma}G^{\rho\sigma} \equiv 0,$$

$$F_{\alpha\beta,\gamma} + F_{\beta\gamma,\alpha} + F_{\gamma\alpha,\beta} \equiv 0$$

sont bien réellement indépendantes. Je suis persuadé que pour pouvoir décider à coup sûr dans tous les cas, il faut avoir une méthode s'appliquant aux équations, *telles qu'elles sont données,* et je suis persuadé que ma méthode s'y applique.

Je m'excuse, cher Maître, de m'être laissé encore aller à un peu de bavardage et vous prie de croire à mes sentiments de haute admiration et d'entier dévouement.

E. Cartan

conditions for involution of this new system are realized, and so on. One is sure to finally finish up in either case 3 or case 2.

The system of 16 equations that I mentioned to you, and of which I write here only a few equations,

$$X_{12} \equiv S_{1,2} - S_{2,1} + a(\phi_{1,2} - \phi_{2,1}) + b(\phi_{3,4} - \phi_{4,3}) - A_{12} = 0,$$
$$Z \equiv \phi_{\underline{\alpha},\alpha} - C = 0,$$
$$Y^{12} \equiv \Lambda^2_{\underline{1\alpha};\alpha} + \Lambda^1_{\underline{2\alpha};\alpha} + c(\phi_{\underline{1,2}} + \phi_{\underline{2,1}}) + d(S_{\underline{1,2}} + S_{\underline{2,1}})$$
$$- g^{12}[\Lambda^\mu_{\mu\underline{\alpha}:\alpha} + \ldots] = 0,$$

falls under the second case; there are 4 linear combinations of the derivatives of the left hand sides in which the second order derivatives cancel; one of these combinations is

$$Y_{\underline{1\alpha};\alpha} - \frac{3c + 1}{2} Z_{;1} - dX_{\underline{1\alpha};\alpha} - \frac{1 + c - ad}{6} h\,(X_{\underline{23};4} + X_{\underline{34};2} + X_{\underline{42};3}).$$

If the quantities A_{12}, C, etc. are arbitrary, the system is inconsistent but one does not know if one can choose them so that the system be completely integrable.

What you say about the number of *real* identities of your system can be reduced, perhaps, to a question of convention. When I say that there are 12 identities, I mean that I use the equations just as they are, without modifying them by the introduction of an auxiliary function. My 12 identities, namely,

$$G^{\alpha\mu}{}_{;\mu} + F^{\alpha\mu}{}_{;\mu} + \Lambda^\rho_{\underline{\alpha\mu}} F_{\mu\rho} \equiv 0,$$
$$G^{\mu\alpha}{}_{;\mu} + \Lambda^\alpha_{\rho\sigma} G^{\rho\sigma} \equiv 0,$$
$$F_{\alpha\beta,\gamma} + F_{\beta\gamma,\alpha} + F_{\gamma\alpha,\beta} \equiv 0$$

are truly independent. I am convinced that in order to be able to definitively decide in all cases one needs a method that can be applied to the equations *as they are given*, and I am convinced that my method does so.

I am sorry, *cher Maître*, to have allowed myself to have run on a little...

E. Cartan

X

Albert Einstein

Berlin W, 18-XII-29
Haberlandstr. 5

Verehrter Herr Cartan!

Ich bin sehr glücklich dass ich Sie als Mit-Strebenden gewonnen habe. Denn Sie haben gerade das, was mir fehlt: eine beneidenswerte Leichtigkeit in der Mathematik. Ihre Ausführungen über den Indice de généralité habe ich noch nicht ordentlich verstanden, wenigstens die Beweise nicht. Ich bitte Sie sehr, mir diejenigen Ihrer Arbeiten zu senden, aus denen ich die Theorie ordentlich studieren kann [1].

Sehr dankbar bin ich Ihnen aber für die Identität

$$G^{\mu\alpha}{}_{;\mu} + \Lambda^{\alpha}_{\rho\sigma} G^{\rho\sigma} \equiv 0,$$

die mir merkwürdigerweise entgangen war. Deshalb habe ich den Umweg über die Funktion ψ nehmen müssen. In einer neuen Darstellung in den *Sitzungsberichten* habe ich von dieser Identität Gebrauch gemacht, indem ich mir erlaubte, auf Sie als Quelle aufmerksam zu machen [2].

Nun zu dem System

$$R_{ik} = 0, \tag{1}$$

$$F_{\alpha\beta} = \phi_{\alpha,\beta} - \phi_{\beta,\alpha} - C(\phi_{\alpha}S_{\beta} - \phi_{\beta}S_{\alpha}) = 0, \tag{2}$$

$$G_{\alpha\beta} = S_{\alpha,\beta} - S_{\beta,\alpha} - C'(\phi_{\alpha}S_{\beta} - \phi_{\beta}S_{\alpha}) = 0. \tag{3}$$

Ich möchte Ihnen sagen, warum ich nicht für wahrscheinlich halte, dass die Feldgesetze der Natur durch diese Gleichungen ausgedrückt werden. (1) enthält *nur* die g_{ik}. Es gäbe also für das reine Gravitationsfeld *allein* eine *determinierte* Gesetzlichkeit, ohne dass die

---

1. On verra la suite à cette demande dans les lettres suivantes.
2. Dans [22] Einstein se réfère à Cartan dans les termes suivants: « La démonstration de la compatibilité est basée sur une communication écrite que je dois à une correspon-

A X

Albert Einstein

Berlin W. 18. XII-29
Haberlandstr. 5

Dear M. Cartan!

I am very fortunate that I have acquired you as a coworker. For you have exactly that which I lack: an enviable facility in mathematics. Your explanation of the *indice de généralité* I have not yet fully understood, at least not the proof. I beg you to send me those of your papers from which I can properly study the theory [1].

But I am very grateful to you for the identities

$$G^{\mu\alpha}{}_{;\mu} + \Lambda^{\alpha}_{\rho\sigma}G^{\rho\sigma} \equiv 0,$$

which, remarkably, had escaped me [2]. It must be for this reason that I had to take the long way round concerning the function ψ. I have made use of these identities in a new article in the *Sitzungsberichten*, in which I took the liberty of citing you as the source.

Now to the system

$$R_{ik} = 0, \tag{1}$$

$$F_{\alpha\beta} = \phi_{\alpha,\beta} - \phi_{\beta,\alpha} - C(\phi_\alpha S_\beta - \phi_\beta S_\alpha) = 0, \tag{2}$$

$$G_{\alpha\beta} = S_{\alpha,\beta} - S_{\beta,\alpha} - C'(\phi_\alpha S_\beta - \phi_\beta S_\alpha) = 0. \tag{3}$$

I should like to tell you why I do not think it likely that the laws of nature are expressed by these equations. (1) contains *only* the g_{ik}. Thus it would provide a *determined* set of laws for the pure gravitational

dance avec M. E. Cartan [voir §3, 16], elle permet une certaine simplification de la présentation faite aux *Math. Annalen*». §3, 16 n'est autre que l'identité en question.

Parallelstruktur darauf zurückwirte. Es gäbe eine kausale Verbindung

$$\text{Metrik} \rightarrow \text{Parallelstruktur}$$

aber keine *umgekehrte* kausale Verknüpfung. Dies ist allzu sonderbar.

Ausserdem ist aber nach Anschauungsweise dies System weniger determiniert als das von mir vorgeschlagene. Denn ich kann (2) und (3) genügen durch das *speziellere* System

$$\phi_\alpha = \frac{\partial \phi}{\partial x^\alpha}, \tag{2'}$$

$$S_\alpha = \frac{\partial S}{\partial x^\alpha}, \tag{3'}$$

wobei ϕ und S zwei Skalare sind. (1), (2') und (3') sind aber 18 Gleichungen für 18 unbekannte Funktionen $h_{\rho\nu}$, ϕ und S. Schreibt man dagegen mein System in der entsprechenden Form, so hat man 20 Gleichungen für 13 Unbekannte, also — wie mir scheint — eine stärkere Determinierung. Oder ist dies falsch gedacht?

Unter Identität habe ich — wie mir scheint — das Gleiche wie Sie verstanden, nachdem ich Ihre ausführliche Definition gelesen habe.

Wenn ich behauptet habe, dass z.B.

$$F_{\alpha\beta,\gamma} + F_{\beta\gamma,\alpha} + F_{\gamma\alpha,\beta} \equiv 0 \quad (\equiv F_{\alpha\beta\gamma})$$

nur 3 unabhängige Identitäten seien, so wollte ich damit sagen, das die Identität

$$0 = F_{\alpha\beta\gamma,\sigma} \delta^{\alpha\beta\gamma\sigma}$$

($\delta^{\alpha\beta\gamma\sigma} = 1$ oder -1 je nach dem Permutations Charakter von $\alpha\beta\gamma\sigma$) immer erfüllt ist, was man auch für $F_{\alpha\beta}$ hineinsetzt.

Die letztere Identität sagt aber nicht darüber uns, wie die $F_{\alpha\beta}$ sich durch die h ausdrücken *. Aehnlich verhält es sich mit meiner dritten Identität.

Unter Verwendung Ihrer Identität habe ich den Kompatibilitätsbeweis so geführt: wir nennen die Identitäten

$$G^{\alpha\mu}_{\;;\mu} + F^{\alpha\mu}_{\;;\mu} + \Lambda^\rho_{\underline{\alpha\mu}} F_{\mu\rho} \equiv 0, \tag{1}$$

$$G^{\alpha\mu}_{\;;\alpha} + \Lambda^\mu_{\rho\sigma} G^{\rho\sigma} \equiv 0, \tag{2}$$

$$F_{\alpha\beta,\gamma} + \dots + \cdots \equiv 0. \tag{3}$$

---

*. Es sind eben nicht *vier unabhängige* Bedingungen für den Bau der $F_{\alpha\beta}(h)$. (A.E.)

field *alone,* without allowing the parallel structure to react upon it. It would give a causal connection of the form

metric → parallel structure

but no *reverse* causal association. This is much too peculiar. And besides, from my point of view, this system is less determined than that offered by me. For I can satisfy (2) and (3) with the *particular* system

$$\phi_\alpha = \frac{\partial \phi}{\partial x^\alpha}, \tag{2'}$$

$$S_\alpha = \frac{\partial S}{\partial x^\alpha}, \tag{3'}$$

where ϕ and S are two scalars. But (1), (2′), and (3′) are 18 equations for 18 unknown functions, h_{sv}, ϕ and S. On the other hand, if one writes my system in the corresponding form one has 20 equations for 13 unknowns, and thus — it appears to me — a stronger determination. Or is this not the case?

After reading your detailed definition, it seems to me that I have the same understanding as you with regard to identities.

When I asserted for example, that

$$F_{\alpha\beta,\gamma} + F_{\beta\gamma,\alpha} + F_{\gamma\alpha,\beta} \equiv 0 \quad (\equiv F_{\alpha\beta\gamma})$$

are only 3 independent identities, I meant that the identity

$$0 = F_{\alpha\beta\gamma,\sigma}\delta^{\alpha\beta\gamma\sigma}$$

($\delta^{\alpha\beta\gamma\sigma} = 1$ or -1 according to the character of the permutation $\alpha\beta\gamma\sigma$) is always satisfied, whatever one takes for $F_{\alpha\beta}$. But this last identity says nothing about how the $F_{\alpha\beta}$ is expressed in terms of the h *. The same thing holds for my third identity.

As for the application of your identities, I have proven their compatibility as follows: let us label the identities as follows

$$G^{\alpha\mu}_{\ ;\mu} + F^{\alpha\mu}_{\ ;\mu} + \Lambda^\rho_{\underline{\alpha}\mu}F_{\mu\rho} \equiv 0, \tag{1}$$

$$G^{\alpha\mu}_{\ ;\alpha} + \Lambda^\mu_{\rho\sigma}G^{\rho\sigma} \equiv 0, \tag{2}$$

$$F_{\alpha\beta,\gamma} + \ldots + \cdots \equiv 0. \tag{3}$$

---

*. There aren't even *four independent* conditions for constructing $F_{\alpha\beta}(h)$. (A.E.)

Sei für den Schnitt $x^4 = a$ eine Lösung aller Gleichungen gefunden. Diese kann dann sicher so stetig fortgesetzt werden, dass die 12 Gleichungen erfüllt sind die durch Null-Setzen von

$$
\begin{array}{ccc}
G^{11} & G^{12} & G^{13} \\
G^{21} & G^{22} & G^{23} \\
G^{31} & G^{32} & G^{33} \\
F_{14} & F_{24} & F_{34}
\end{array}
$$

entstehen. Aus den letzten drei in Verbindung mit (3) folgt, dass auch $\dfrac{\partial F_{23}}{\partial x^4}$, $\dfrac{\partial F_{31}}{\partial x^4}$ und $\dfrac{\partial F_{12}}{\partial x^4}$ überall verschwinden. Es verschwinden also (weil auch in $x^4 = a$) die linken Seiten F_{23}, etc. überall.

Aus (1) und (2) folgt nun, dass in $x^4 = a$ auch die Ableitungen nach x^4 von G^{14}, G^{41}, ..., G^{44} verschwinden; also verschwinden diese Grössen in $x^4 = a + da$. Durch Fortsetzung folgt die Erfüllung dieser Gleichungen im ganzen Gebiete. Wahrscheinlich machen Ihre allgemeinen Methoden eine solche Demonstration überflüssig; ich aber habe mir *so* geholfen.

Noch eine Bemerkung. Die Gleichungen

$$
\begin{cases}
(\Lambda^\alpha_{\mu\nu;\sigma} + \ldots + \ldots) + (\Lambda^\alpha_{\mu\rho}\Lambda^\rho_{\nu\sigma} + \ldots + \ldots) = 0, \\
\Lambda^\alpha_{\underline{\mu\nu};\nu} - \Lambda^\rho_{\underline{\mu\tau}}\Lambda^\alpha_{\sigma\tau} = 0, \\
\Lambda^\alpha_{\mu\nu;\alpha} = 0,
\end{cases}
$$

haben die hässliche Eigenschaft ihren quadratischen Gliedern die $\varDelta$ zu enthalten (bei Ausführung von ;). Es ist doch sonderbar, dass es kein System geben soll, in dem *nur* die $\varLambda$ vorkommen. So ist das System nicht „ kanonisch ". Allerdings kann man die *h* als Feldvariable zufügen und setzen

$$
h_{s\mu,\nu} = h_{s\alpha}\,\varDelta^\alpha_{\mu\nu}.
$$

Betrachtet man nun die *h* und $\varDelta$ als Feldvariable, dann treten Ableitungen nur linear auf. Aber dies ist hässlich, weil die $\varDelta$ nicht Tensoren sind.

Herzlich grüsst Sie

Ihr

A. Einstein

Let a solution of all the equations have been found on a cross-section $x^4 = a$. This can certainly be extended in so continuous a manner that the 12 equations arising from setting

$$
\begin{array}{ccc}
G^{11} & G^{12} & G^{13} \\
G^{21} & G^{22} & G^{23} \\
G^{31} & G^{32} & G^{33} \\
F_{14} & F_{24} & F_{34}
\end{array}
$$

equal to zero are all satisfied. From the last three, together with (3), it follows that $\dfrac{\partial F_{23}}{\partial x^4}$, $\dfrac{\partial F_{31}}{\partial x^4}$ and $\dfrac{\partial F_{12}}{\partial x^4}$ also vanish everywhere. Thus the left hand sides, F_{23}, etc., vanish everywhere (since they do on $x^4 = a$).

Now, it follows from (1) and (2) that on $x^4 = a$ the derivatives of G^{14}, G^{41}, ..., G^{44} with respect to x^4 vanish; hence, these quantities vanish on $x^4 = a + da$. That these equations are satisfied in the whole domain follows from continuation. Your general methods probably make such a demonstration superfluous; but *this* is the way that is helpful for me.

One more remark. The equations

$$
\left\{
\begin{array}{l}
(\Lambda^{\alpha}_{\mu\nu;\sigma} + \ldots + \ldots) + (\Lambda^{\alpha}_{\mu\rho}\Lambda^{\rho}_{\nu\sigma} + \ldots + \ldots) = 0, \\
\Lambda^{\alpha}_{\underline{\mu\nu};\nu} - \Lambda^{\rho}_{\underline{\mu}\tau}\Lambda^{\alpha}_{\sigma\tau} = 0, \\
\Lambda^{\alpha}_{\mu\nu;\alpha} = 0,
\end{array}
\right.
$$

have the unpleasant property that they contain Δ in their quadratic terms (in the expression for ;). Surely it is strange that there should be no system in which *only* the Λ appear. Hence the system is not " canonical ". Indeed, one can add on the h as a field variable and set

$$
h_{s\mu,\nu} = h_{s\alpha}\Delta^{\alpha}_{\mu\nu}.
$$

If one now considers the h and Δ as field variables, then the derivatives arise only linearly. But this is unpleasant because the Δ are not tensors.

Kind regards.

Yours,

A. Einstein

Am wichtigsten ist mir die Frage: Halten Sie mein Gleichungssystem vom Gesichtspunkte der Stärke der Determination für privilegiert? Denn erst, wenn man dies mit Recht behaupten darf, erhält die ganze Theorie etwas Zwingendes.

The most important thing for me is the question: from the point of view of the degree of their determination, do you consider my system of equations to be especially privileged? For only if one may correctly assert this does the whole theory acquire any cogency.

XI

Le Chesnay (Seine et Oise)
27 avenue de Montespan,
le 22 décembre 1929

Cher et illustre Maître,

Je voudrais vous éviter la lecture des mémoires où j'ai exposé ma théorie des systèmes en involution; ils sont bien longs; de plus ils considèrent les systèmes différentiels sous la forme d'équations aux différentielles totales, tandis que c'est sous la forme d'équations aux dérivées partielles que se présentent les systèmes qui vous intéressent. Je crois que je puis vous donner un exposé suffisamment précis, avec les démonstrations, en un petit nombre de pages (dans la note que je vous avais envoyée il n'y a pas de démonstration). Dans très peu de jours je pense vous envoyer cette nouvelle note.

Venons maintenant aux points soulevés par votre lettre. Je n'ai pas très bien compris ce que vous me dites au sujet des raisons qui rendent peu vraisemblable le système avec $R_{ik} = 0$. Dans la phrase: « *Es gäbe also für das reine Gravitationsfeld* ALLEIN *eine* DETERMINIERTE *Gesetzlichkeit, ohne dass die Parallelstruktur darauf zurückwirkte* », le mot *allein* se rapporte-t-il à « *Gravitationsfeld* » ou à « *eine determinierte* etc... »? Je ne suis pas assez fort en allemand pour le décider. Je ne comprends pas non plus très bien quelle est cette « *Kausale Verbindung Metrik → Parallelstruktur* ». Il est certain que les 10 équations $R_{ik} = 0$ ne font intervenir que la métrique. Mais ne pourrait-on pas dire aussi que les 6 équations $\Lambda^{\mu}_{\alpha\beta;\mu} = 0$ de votre système ne font intervenir que le parallélisme? Du reste je ne vous ai signalé ce système que comme une solution possible du problème *mathématique* posé.

Je ne suis pas d'accord avec vous sur le plus ou moins de détermination des deux systèmes de 22 équations, celui qui contient les $R_{ik} = 0$ et le vôtre. Ils ont certainement le même degré de détermination. Votre système peut se mettre sous une forme tout à fait analogue à l'autre, à savoir

$$\phi_\alpha = \frac{1}{\psi} \frac{\partial \psi}{\partial x^\alpha}, \tag{1}$$

A XI

Le Chesnay (Seine et Oise)
27 avenue de Montespan,
22 December 1929

Cher et illustre Maître,

I should like to spare you the trouble of reading the papers where I have outlined my theory of systems in involution since they are quite long. Moreover, they consider differential systems in the form of total differential equations whereas the systems you are interested in appear in the form of partial differential equations. I think I can give you a sufficiently precise account, with proofs, in a few pages (In the note that I sent to you, there are no proofs). I expect to send you this new note in a very few days.

Let us consider the points raised in your letter. I have not fully understood what you say about the reasons that render the system with $R_{ik} = 0$ not very likely. In the sentence: " *Es gäbe also für das reine Gravitationsfeld* ALLEIN *eine* DETERMINIERTE *Gesetzlichkeit, ohne dass die Parallelstruktur darauf zurückwirkte* ", does the word *allein* refer to " *Gravitationsfeld* " or to " *eine determinierte* etc " ? My German is not good enough to decide this. Nor do I quite understand the point of this " *Kausale Verbindung Metrik → Parallelstruktur* ". It is clear that the 10 equations $R_{ik} = 0$ involve only the metric. But can't we also say that the 6 equations $\Lambda^{\mu}_{\alpha\beta;\mu} = 0$ of your system involve only the parallelism? Besides, I mentioned this system only as a possible solution to the *mathematical* problem.

I do not agree with you on the greater or lesser determination of the two systems of 22 equations, the one that contain $R_{ik} = 0$ and yours. They certainly have the same degree of determinaion. Your system can be put in a form quite similar to the other, namely

$$\phi_\alpha = \frac{1}{\psi} \frac{\partial \psi}{\partial x^\alpha}, \tag{1}$$

$$S_\alpha = -\frac{1}{\psi}\frac{\partial \chi}{\partial x^\alpha}, \tag{2}$$

$$G^{\alpha\beta} + G^{\beta\alpha} = 0, \tag{3}$$

avec deux scalaires ψ et χ, ce qui fait encore 18 équations pour 18 fonctions inconnues h_{sv}, ψ et χ. Les équations (1) et (2) résultent de l'intégration des équations

$$F^{\alpha\beta} = 0, \ G^{\alpha\beta} - G^{\beta\alpha} = 0;$$

on a en effet, par exemple,

$$h(G^{34} - G^{43} - F^{34}) \equiv \frac{\partial S_1}{\partial x^2} - \frac{\partial S_2}{\partial x^1} + \phi_1 S_2 - \phi_2 S_1.$$

Les deux systèmes ont donc une structure tout à fait analogue, et leurs deux indices de généralités sont les mêmes.

On ne peut donc pas dire que votre système soit privilégié en ce qui concerne le degré de détermination; néanmoins je crois pouvoir affirmer, après avoir formé toutes les identités possibles et passé en revue les différents systèmes susceptibles de convenir, qu'*il n'existe pas de système* PLUS *déterminé que le vôtre*.

J'ai maintenant parfaitement compris ce que vous entendiez en disant que les 4 identités $F_{\alpha\beta\gamma} = 0$ ne sont pas indépendantes. C'est que nous ne donnons pas au mot *indépendantes* la même signification. Pour moi il s'agissait d'indépendance *algébrique* entre les premiers membres des 4 identités, considérées comme combinaisons linéaires des 24 dérivées partielles des quantités $F_{\alpha\beta}$ par rapport à x^1, x^2, x^3, x^4. Notre point de vue était donc tout à fait différent.

J'avais pensé un moment que, toutes les fois qu'il y aurait $N - 12 + r_2 > N - 12$ identités, il existerait, entre ces identités, r_2 relations identiques à votre sens. Mais il n'en est rien et je suis persuadé que la non-indépendance (à votre sens) des identités ne joue aucun rôle dans la question de compatibilité. Et cela est fort heureux, sinon en effet vous seriez obligé, quand vous avez trouvé les $N - 12$ identités nécessaires, de démontrer qu'elles sont indépendantes *à votre sens*, ce qui ne serait pas toujours facile.

Je ne vous parle pas aujourd'hui de votre nouvelle manière de présenter la démonstration de la compatibilité; elle est du reste fort analogue à la manière dont je démontre l'existence des solutions de mes systèmes en involution, mais je crois qu'il ne suffit pas de démon-

$$S_\alpha = -\frac{1}{\psi}\frac{\partial\chi}{\partial x^\alpha}, \tag{2}$$

$$G^{\alpha\beta} + G^{\beta\alpha} = 0, \tag{3}$$

with two scalars ψ and χ, and this again gives 18 equations for the 18 unknown functions h_{sv}, ψ and χ. Equations (1) and (2) follow from integrating the equations

$$F^{\alpha\beta} = 0, \; G^{\alpha\beta} - G^{\beta\alpha} = 0.$$

Indeed, as an example one has

$$h(G^{34} - G^{43} - F^{34}) \equiv \frac{\partial S_1}{\partial x^2} - \frac{\partial S_2}{\partial x^1} + \phi_1 S_2 - \phi_2 S_1.$$

So, the two systems have a quite similar structure, and their two generality indexes are equal.

Therefore, one cannot say that your system is more privileged as far as the degree of determination is concerned; nevertheless, I think I can assert, having written down all the possible identities and reviewed the different systems likely to fit, *that there is no system* MORE *determined* than yours.

I have now understood perfectly what you mean when you say that the 4 identities $F_{\alpha\beta\gamma} = 0$ are not independent. We do not give the same meaning to the word *independent*. In any case, I was concerned with *algebraic* independence among the left hand sides of the 4 identities, considered as linear combinations of the 24 partial derivatives of the quantities $F_{\alpha\beta}$ with respect to x^1, x^2, x^3, x^4. Our points of view were thus completely different.

I thought for a while that each time there are $N - 12 + r_2 > N - 12$ identities, r_2 relations, identical in your sense would exist between these identities. But nothing of the kind occurs and I am convinced that the non independence (in your sense) of the identities does not play any role in the question of compatibility. And this is quite a lucky thing since otherwise you would be obliged, whenever you found the necessary $N - 12$ identities, to prove that they were independent *in your sense*, and this would not always be easy.

I won't mention today your new way of presenting the proof of compatibility. In any case, it is very similar to the way I prove the existence of the solutions of my systems in involution, but I think that

trer l'existence d'une solution dans tout l'espace en partant d'une solution, dans la section $x^4 = a$, du système partiel

$$G^{11} \quad G^{12} \quad G^{13}$$
....

Il faut encore démontrer la compatibilité de ce système partiel. Cela ramène à mon point de vue, à savoir la considération des solutions à 1, 2 et 3 dimensions.

Je ne vous parle pas non plus du système à 16 + 22 équations qui vous paraît « *hässlich* ». Je crois aussi que je pourrai vous présenter d'une manière élémentaire — bien qu'à mon sens pas absolument satisfaisante — le problème du degré de généralité des espaces riemanniens à parallélisme absolu, considérés dans ce qu'ils ont d'*essentiel*, c'est-à-dire d'indépendant du choix des variables.

Puis-je maintenant vous demander ce que vous pensez des considérations que j'ai exposées au début de ma note sur le déterminisme *intégral* (au sens de l'ancienne mécanique) et le déterminisme *local*. Cette question est liée à la suivante.

Toute solution des équations de Maxwell définie seulement dans une petite région de l'espace-temps est-elle *physiquement* admissible? Autrement dit supposons que nous ayons pour

$$|x - x_0| < a, \ |y - y_0| < b, \ |z - z_0| < c, \ |t - t_0| < h$$

des fonctions définissant dans ce domaine le champ électromagnétique et satisfaisant, *dans ce domaine limité*, aux équations de Maxwell; pensez-vous qu'un tel champ puisse se présenter en réalité? L'état local est certainement solidaire de ce qui peut se passer ailleurs, en dehors du domaine considéré, mais peut-on disposer des possibilités, en nombre infini, qui existent à l'extérieur du domaine pour rendre possible, à l'intérieur du domaine, n'importe quelle solution *locale* des équations de Maxwell?

Cette question me semble importante parce que, si elle était résolue par l'affirmative, cela donnerait un sens *physique* et non plus seulement mathématique à l'indice de généralité du système d'équations aux dérivées partielles qui régit le champ.

Mais en voilà assez pour aujourd'hui. Je vous prie, cher et illustre Maître, de vouloir bien agréer l'hommage de mes sentiments d'admiration et de cordial dévouement.

E. Cartan

it's not enough to prove the existence of a solution in a whole space starting from a solution in the section $x^4 = a$ of partial system

$$G^{11} \quad G^{12} \quad G^{13}$$
....

It is still necessary to prove the compatibility of this partial system. This brings us back to my point of view, namely, the consideration of the 1, 2, and 3-dimensional solutions.

Nor will I mention the system of 16 + 22 equations that seems " *hässlich* " to you. This, too, I think I shall be able to present to you in an elementary way — though not in my opinion an entirely satisfying one — the problem of the generality degree of Riemannian spaces with absolute parallelism, with regard to their *essential* properties, that is, the properties independent of the choice of variables.

Allow me to ask you now what you think of the ideas I put forward, at the beginning of my note, about *global* determinism (in the sense of the old mechanics) and *local* determinism. This question is related to the following.

Is any solution of Maxwell's equations which is defined only in a small region of space-time, *physically* admissible? To put it another way, let us assume that for

$$|x - x_0| < a, \; |y - y_o| < b, \; |z - z_o| < c, \; |t - t_0| < h$$

we have functions defining the electromagnetic field in this domain and satisfying Maxwell's equations *in this limited domain*; do you think that such a field can actually exist? The local state is certainly influenced by what happens elsewhere, outside the domain under consideration, but can we deal with the possibilities, infinite in number, that exist outside the domain so as to determine any *local* solution of Maxwell's equations inside the domain?

This question seems important because, if the answer is affirmative, it would give a *physical* meaning and not merely a mathematical one, to the generality index of the system of partial differential equations that determines the field.

But that is enough for to-day...

E. Cartan

XII

Le Chesnay, le 25.12.1929

Cher et illustre Maître,

Je vous envoie la seconde note où vous trouverez la théorie complète des systèmes en involution, avec les démonstrations; je l'ai rédigée en me plaçant au point de vue des systèmes d'équations aux dérivées partielles et non, comme dans mes mémoires, au point de vue des systèmes d'équations aux différentielles totales. J'ai indiqué quelques exemples [1].

Si vous pouvez aller jusqu'à la fin sans trop de difficulté, je vous demanderai ce que vous pensez du dernier alinéa page 17 [2].

Je profite de l'occasion, cher et illustre Maître, pour vous adresser mes meilleurs vœux de nouvel an, vœux de santé pour vous et vœux pour une riche moisson dans votre nouvelle théorie du champ. Votre bien dévoué

E. Cartan

---

1. Le manuscrit inédit de Cartan de 17 pages in-4° est très proche de [13] quant au contenu et même à la forme du texte et au choix des exemples.

Comme dans le texte publié Cartan aborde successivement la discussion des systèmes en involution à 2, 3 et à un nombre quelconque de variables indépendantes pour se terminer par l'application à la théorie unitaire du champ. Dans [13] Cartan entre dans plus de détails que dans le manuscrit inédit et il donne aussi les variantes aux équations de champ d'Einstein, celles à 22 et 15 équations proposées dès la lettre VII.

2. La remarque est la suivante: dans la théorie classique, fluide gravitant sans tension,

A XII

25 December 1929

Cher et illustre Maître,

I am sending you the second note, in which you will find the complete theory of systems in involution, with proofs. It is written from the point of view of systems of partial differential equations and not, as in my papers, from the point of view of systems of total differentials. I have included a few examples [1].

If you manage to reach the end without too many problems, tell me what you think of the last paragraph on page 17 [2].

I am taking this opportunity, *cher et illustre Maître*, to send you my best wishes for the New Year, wishes for your health and for a rich harvest of results from your new unified field theory.

E. Cartan

---

les sections à 3 dimensions d'espace-temps obtenues en suivant la surface du fluide dans son mouvement sont des caractéristiques des équations du champ. Ici, je cite: « Ces caractéristiques ne correspondent à rien dans la théorie projetée du champ gravitationnel électromagnétique. Cela ne prouverait-il pas a priori que dans la nouvelle théorie la matière ou l'électricité, du moins à l'état continu, est un tout dont il est impossible d'individualiser les éléments, ce qui conduirait à la négation de la substance? ».

On verra qu'effectivement Einstein achoppera à cette difficulté pour abandonner plus tard sa tentative.

XIII

[27 ou 28-XII-29]

Vereherter College!

Gleich nach Absendung meines Briefes wurde mir schon klar, dass vom Standpunkte des Determinations-Grades Ihr System $R_{ik} = 0$ etc. dem meinigen völlig äquivalent sei. Ich hatte vergessen, dass man auch meinem System eine analoge Form geben kann (mit ψ und χ), sodass nur mehr vier Identitäten auftreten.

Dagegen ist mein anderes Bedenken nach meiner Ansicht stichhaltig. Die Gleichungen $R_{ik} = 0$ allein bestimmen das g_{ik} Feld vollständig, und es spielt bei dieser Bestimmung die Parallelstruktur keine Rolle. Sind in einem Zeitschnitte $x^4 =$ konst. die g und ihre ersten Ableitungen gegeben, so ist die analytische Fortsetzung der g (bis auf die Willkür, welche der Freiheit der Koordinatenwahl entspricht) vollständig bestimmt. D.h. die g_{ik} allein haben ihre selbständige Kausalität, unabhängig von der Parallelstruktur. Man kann es auch so ausdrücken: das elektromagnetische Feld hat keine Rückwirkung auf das Gravitationsfeld. Dies widerspricht der physikalischen Erwartung durchaus.

Analoges gilt nicht für die Gleichungen $\Lambda^\alpha_{\mu\nu;\alpha} = 0$ meines Systems bezüglich der Parallelstruktur. Die Parallelstruktur lässt sich überhaupt nicht aus der Gesamtheit der Variabeln heraussondern, d.h. es gibt nicht eine Gesamtheit von 6 Grössen, die sich aus den h algebraisch ausdrücken, derart, dass eine Analogie zu den Grössen

$$g_{\mu\nu} = h_{s\mu}h_{s\nu}$$

geschaffen wäre. In den Gleichungen $\Lambda^\alpha_{\mu\nu;\alpha} = 0$ treten nicht ausschliesslich bestimmte Kombinationen der h auf, deren Kausalität durch diese Gleichungen *allein* bestimmt würde.

Allerdings erweckt die Auffindung Ihres Systems den Verdacht, dass es vielleicht noch andere kompatible Gleichungssysteme von demselben Determinationsgrad geben könnte wie meines, und welche den Schönheitsfehler der Gleichungen $R_{ik} = 0$ *nicht* haben. Dann hätte ich kein Recht mehr, meine Gleichungen für privilegiert zu halten. Wie denken Sie darüber?

A XIII

[27 or 28-12-29]

Dear Colleague,

Immediately after I sent off my letter it became clear to me that, from the standpoint of the degree of determination, your system, $R_{ik} = 0$, etc., is fully equivalent to mine. I had forgotten that my system too can be given an analogous form (with ψ and χ), such that now four identities appear.

On the other hand, I think that my other objection is sound. The equations $R_{ik} = 0$ alone fully determine the g_{ik}-field, and in this the parallel structure plays no role. If the g and their first derivatives are given on a space-like cross-section, $x^4 = const.$, then the analytic continuation of the g's is fully determined (up to the arbitrariness corresponding to the freedom in the choice of coordinates). That is, the g_{ik} alone have an autonomous causality, independent of the parallel structure. This can be expressed as follows: the electromagnetic field has no reaction upon the gravitational field. This completely contradicts physical expectations.

The same thing does not hold true for the equations $\Lambda^\alpha_{\mu v;\alpha} = 0$ of my system with respect to the parallel structure. The parallel structure can by no means be separated out from the whole collection of variables, i.e. there does not exist a total of 6 quantities which can be algebraically expressed in terms of the h so as to produce an analogy to the quantities

$$g_{\mu v} = h_{s\mu} h_{sv}.$$

In the equations $\Lambda^\alpha_{\mu v;\alpha} = 0$, no uniquely determined combination of the h appears whose causality would be fixed by these equations *alone*.

Of course, the discovery of your system awakens the suspicion that, perhaps, yet another compatible system of equations can be given having the same degree of determination as mine but which do *not* have the aesthetic drawback of the equations $R_{ik} = 0$. In that case, I could no longer consider my equations to be privileged. What do you think about this?

Es wäre eine schöne Sache, wenn Sie eine Abhandlung über den Determinationsgrad von partiellen Differentialgleichungen drucken lassen würden. Das könnte doch auch die Mathematiker interessieren. Ich bin aus Ihrer schriftlichen Darlegung nicht vollkommen klug geworden. Besonders macht es mir Schwierigkeiten, dass die Gleichungen — wenn man die Λ und h als Feldwariable nimmt, in den quadratischen Termen die h in differenzierter Form enthalten, was doch — wir mir scheint — Ihre allgemeine Theorie nicht voraussetzt.

Ich freue mich darüber, dass nach Ihren Untersuchungen stärker determinierte kompatible Gleichungssysteme als unsere jetzigen nicht existieren. Bewundernswert, dass Sie so etwas beweisen können!

Sie haben natürlich Recht, wenn Sie sagen, dass ich die Erfüllbarkeit aller Gleichungen für $x^4 = a$ nicht bewiesen sondern nur angenommen habe. Aber wenn wir Physiker keine schlimmeren mathematischen Sünden begingen als diese, so wären wir verhältnismässig brave Leute! Ich bin glücklich, dass es mir gelungen ist, Sie für diese Sachen zu interessieren und wusste, dass Sie viel Klarheit hineinbringen werden.

Ihre Frage über die Fortsetzbarkeit der Lösungen von Differentialgleichungen von *physikalischen Standpunkt aus* ist mir nicht ganz klar. Ich versuche aber zu antworten so gut ich kann.

Es scheint mir darauf anzukommen was man vom physikalischen Standpunt aus von der Lösung verlangt.

Wenn wir z.B. die Maxwell'schen Gleichungen des leeres Raumes betrachten, so ist ja wegen des hyperbolischen Charakters die Fortsetzung nach den räumlichen Dimensionen nicht bestimmt (Wellen aus dem Unendlichen!). Davon sehe ich aber gleich ab, weil es nicht das Wesentliche Ihrer Frage betrifft. Ich beschränke mich also z.B. auf statische Probleme.

Ist nun jede Lösung im endlichen Gebiet sinnvoll fortsetzbar? Dies hängt davon ab, *was für sonstige Eigenschaften wir von dem Felde aus physikalischen Gründen verlangen zu müssen glauben* (*Erlaubtheit von Singularitäten*, oder von Grenzbedingungen gewisser Art im Endlichen oder Unendlichen). Bei den Maxwell'schen Gleichungen gibt es z.B. keine singularitätsfreien Lösungen, die im Unendlichen nach null tendieren. Hat man eine Lösung in Ihrem Gebiete, so wird man bei der Fortsetzung zu Singularitäten oder Wachsen der Felder im Unendlichen gelangen. Die Zulässigkeit Ihrer Lösungen im Endlichen wird davon abhängen, welche derartige " Extravaganzen " der

It would be a good thing if you were to publish an article on the degree of determination of partial differential equations; this could also be of interest to mathematicians. I have not been entirely enlightened by your written explanation. I have especial difficulties with the fact that the equations — when one takes Λ and h as field variables — contain h in its differential form in the quadratic terms; something — it seems to me — that your general theory does not deal with.

I am delighted that, according to your investigations, more strongly determined, compatible systems of equations than our current ones do not exist. It's wonderful that you can prove such a thing.

You are, of course, right when you say that I have not proven, but only assumed, that all the equations can be satisfied at $x^4 = a$. But if we physicists committed no worse mathematical sins than this we would be comparatively honest folk! I am lucky that I succeeded in interesting you in such matters, and I knew that you would bring much clarity to them.

Your question concerning the continuability of the solutions of differential equations *from the physical standpoint* is not totally clear to me. But I shall try to answer as best I can.

It seems to me to be necessary to look at what one demands of a solution from the physical point of view.

For example, if we look at Maxwell's equations for empty space, then, because of their hyperbolic character, continuation of a solution in spacelike directions is not determined (waves coming in from infinity!). But I leave this aside because it does not touch upon the essence of your question. Therefore, I restrict myself to, say, static problems.

Now, is every solution meaningfully continuable in a finite region? This depends on *what other properties of the field we feel we must demand on physical grounds* (the admissibility of singularities, or finite or infinite boundary conditions of a certain kind). In the case of Maxwell's equations, for example, there are no singularity-free solutions which tend to zero at infinity. If you have a solution in your neighbourhood, then, by continuation, you would arrive at a singularity or at growth of the field at infinity. The admissibility of your solution in a finite region then depends on what kind of " extravagances " of the solution one believes to be permissible from the physical point of view.

Lösung man von physikalischen Standpunkte aus zulassen zu dürfen glaubt.

Wie schon gesagt, kommt man bei den Maxwell'schen Gleichungen nicht ohne Singularitäten aus. Aber kein vernünftiger Mensch glaubt, dass die Maxwell'schen Gleichungen strenge gültig sein können. Sie sind günstigen Falles erste Approximationen für schwache Felder.

Es ist nun meine Überzeugung, dass bei ernst zu nehmenden, strengen Feldtheorien *völlig Singularitätsfreiheit des ganzen Feldes* verlangt werden muss. Dies wird wohl die freie Wahl der Lösungen in einem Gebiete in einer sehr weitgehenden Weise einschränken — über die Einschränkungen hinaus, die Ihren Determinationsgrade entsprechen. Dessenungeachtet halte ich die Theorie des letzteren für sehr wichtig und klärend. Eine ausfürhliche Publikation darüber aus Ihrer Feder wäre sicher für alle wertwoll, die sich für prinzipielle Fragen interessieren.

Indem ich Ihnen aufs Neue herzlich danke für die viele Mühe, die Sie sich mit der Sache und mit — mir geben, bin ich mit herzlichen Grüssen

Ihr dankbarer

A. Einstein

As I've said, one doesn't get away without singularities in the case of Maxwell's equations. But no reasonable person believes that Maxwell's equations can hold rigorously. They are, in suitable cases, first approximations for weak fields.

It is now my belief that, for a serious and rigorous field theory, one must insist that *the field be free of singularities everywhere.* This will probably restrict the free choice of solution in a region in a very severe way — over and above the restrictions which correspond to your degree of determination. Nevertheless, I think the theory of this last is very clarifying and important. A detailed article on this from your pen would certainly be useful to everyone who is interested in these as fundamental issues.

Again, many thanks for all the pains you have taken with the research — and with me. With kind regards,

Gratefully yours,

A. Einstein

XIV

[29 ou 30-XII-29]

Verehrter Herr Cartan!

Ich habe Ihr Manuskript gelesen und zwar *mit Begeisterung*. Nun ist mir alles klar. Sie sollten diese Theorie ausführlich publizieren, denn ich glaube, dass sie fundamentale Bedeutung hat. Ich hatte mit meinem Assistenten Prof. Müntz früher etwas Aehnliches versucht — wir sind aber nicht durchgekommen. Im Folgenden nur einige Bemerkungen.

Zuerst Ihre Frage wegen der Nichtexistenz einer " Substanz ". Substanz in Ihrem Sinne bedeutet Existenz von zeitartigen Linien besonderen Art. Es ist dies die Übertragung des Teilchen-Begriffes auf das Kontinuum. Eine solche Übertragung bezw. die Notwendigkeit einer Übertragbarkeit auf das Kontinuum als eine theoretische Forderung erscheint ganz unberechtigt. Um das Wesentliche des Gedankens der Atomistik auf dem Boden der Kontinuums-Theorie zu realisieren, genügt es räumliche kleine Gebiete hoher Feldstärke zu haben, die bezüglich ihrer " zeitlichen " Fortsetzung gewissen integralen Erhaltungssätzen genügen. D.h. der *ganze* Komplex muss eine Art von individueller Fortsetzung haben, nicht aber seine einzelnen *Punkte* bezw. Teile.

Ihre wunderbaren Darlegungen lassen für mich noch folgende Fragestellungen übrig.

1) Gibt es Systeme mit nicht trivialen Lösungen, deren " *indice de généralité* " = 0 ist (abgesehen von der Folge der Koordinaten-Willkür)? Dies wären Systeme, welche denen der klassichen Mechanik weitgehend analog wären, indem die Anfangsbedingungen nicht durch eine Anzahl willkürliche vorzugebender *Funktionen* sondern etwa durch eine Anzahl Parameter (Zahlen) bestimmt wären.

Eine Zeitlang war ich überzeugt, dass die wahren Naturgesetze von solcher Art sein müssten.

2) Es wäre aber auch möglich, dass jener hohe Grad von Gebundenheit, über dessen Realisierung in den wahren Naturgesetzen

A XIV

[29 or 30-12-29]

Dear M. Cartan,

I have read your manuscript — *enthusiastically*. Now everything is clear to me. You should publish this theory in detail; I believe it is of fundamental importance. Previously, my assistant, Prof. Müntz and I had sought something similar — but we were unsuccessful. Just a few remarks.

First, your question concerning the nonexistence of a " substance ". Substance, in your sense, means the existence of timelike lines of a special kind. This is the translation of the concept of particle to the case of a continuum. Such a translation into the continuum, or the necessity of such translatability, seems totally unreasonable as a theoretical demand. To realise the essential point of atomistic thought on the level of continuum theory, it is sufficient to have a field of high intensity in a spatially small region which, with respect to its " timelike " evolution, satisfies certain integral conservation laws; i.e., the *whole* complex must have a kind of individual evolution, but not its individual *points* or pieces.

Your wonderful explanations leave me only the following few questions:

1) Are there systems with nontrivial solutions whose " *indice de généralité* " = 0 (apart from the consequences of coordinate freedom)? These would be systems which would be largely analogous to those of classical mechanics, in which the initial conditions would be fixed not by an arbitrary choice of given *functions* but by a choice of parameters (numbers).

For some time I was convinced that the true laws of nature would have to be of such a kind.

2) But it might also be possible that higher degree of constraint (which I have no doubt is realized in the true laws of nature) is based on something else.

für mich kein Zweifel besteht, auf etwas anderem beruht. Das angedeutete geringe Mass von Willkür (in der Natur) könnte auch darin seinen Grund haben, dass Singularitäten im ganzen Raume auszuschliessen sind.

Meine Gleichungen könnten nur dann Aussicht auf Gültigkeit haben, wenn sie *singularitätsfreie* Lösungen haben, welche die materiellen Ladungen darzustellen geeignet sind.

Gibt es wohl allgemeine Methoden, um über die Existenz und Beschaffenheit solcher Lösungen etwas zu erfahren?

3) Die Aufklärung *eines* Punktes wäre nach meiner Ansicht noch sehr wünschenswert. Wenn wir ein System haben, das *nicht* in Involution ist, indem nicht die nötigen Identitäten bestehen

$$\text{z.B.} \quad A^{\alpha}_{\mu\nu;\nu} = 0,$$

was kann man dann über die Lösungen eines solchen Systems aussagen?

In diesem Falle sind Ihre Gleichungen

$$\frac{\partial F_i}{\partial y} - \sum_j A_{ij} \frac{\partial G_j}{\partial x} = 0 \tag{4}$$

neue Bedingungen für die *Fortsetzung* (Bezw. für den Schnitt $x^4 = a$). Kann man diese Vermehrung der Bedingungen für die Fortsetzung dadurch immer weiter vermehren, dass man höher differenziert und in passender Weise gewisse Differenzialquotienten eliminiert? Wie kann man sich über den Allgemeinheitsgrad der Lösungen eines solchen Systems irgend eine Vorstellung machen?

Ich selbst hielt zuerst solche Gleichungen als Ausdruck von Naturgesetzen für möglich. Diese Meinung korrigierte ich aber durch eine Betrachtung, die sich auf die erste Näherung bezog.

Seien G = 0 (1) die Feldgleichungen. Wie zerlegen sie in

$$\overline{G} + \overline{\overline{G}} = 0, \tag{1a}$$

wobei die $\overline{G}$ alle Glieder umfassen, welche in den $h_{s\nu} - \delta_{s\nu} = \bar{h}_{s\nu}$ *linear* sind, während die $\overline{\overline{G}}$ in den $\bar{h}_{s\nu}$ von höheren Grade sind. Seien $L(\overline{G}) \equiv 0$ gewisse lineare Identitäten (d.h. L lineare Differenzialausdrücke). Dann folgt aus den Feldgleichungen

$$L(\overline{\overline{G}}) = 0. \tag{2}$$

The indicated small measure of arbitrariness (in Nature) could also be grounded in the requirement that singularities be excluded from all space!

My equations, then, would only have a possibility of validity if they were to possess *singularity-free* solutions appropriate for the representation of material charges.

Can there possibly be general methods by which we may learn something about the existence and properties of such solutions?

3) In my opinion, the explanation of *one* point would still be desirable: if we have a system that is *not* in involution, in which the necessary identities,

$$\text{e.g. } A^{\alpha}_{\mu y;\nu} = 0,$$

do *not* hold, then what can one say about the solutions of such a system?

In this case, your equations

$$\frac{\partial F_i}{\partial y} - \sum_j A_{ij} \frac{\partial G_j}{\partial x} = 0 \tag{4}$$

are new conditions for *continuation* (or for the cross-section $x^4 = a$). Can one always further augment these additional conditions for continuation by differentiating up and, in certain cases, eliminating given derivatives? How can one get some kind of idea about the degree of generality of the solutions of such a system?

At first, I myself though such equations possible as expressions of laws of nature. But I changed my mind by a study of the first approximation.

Let

$$G = 0 \tag{1}$$

be the field equations. We break them up into

$$\overline{G} + \overline{\overline{G}} = 0 \tag{1a}$$

where $\overline{G}$ represents all the terms which are *linear* in $h_{sy} - \delta_{sy} = \overline{h}_{sy}$, while $\overline{\overline{G}}$ is of higher order in $\overline{h}_{sy}$. Let $L(\overline{G}) \equiv 0$ be certain linear identities (i.e., L are linear differential expressions). Then it follows from the field equations that

$$L(\overline{\overline{G}}) = 0. \tag{2}$$

If the linear identities

$$L(\overline{G}) \equiv 0$$

Wenn den linearen Identitäten $L(\overline{G}) \equiv 0$ keine strengen Identitäten entsprechen, so ergibt sich folgender merkwürdige Umstand.

Man kann eine strenge Lösung der Gleichungen so entwickeln

$$h_{sv} = \delta_{sv} + \overline{h}_{sv} + \overline{\overline{h}}_{sv} + \ldots,$$

was eine Entwicklung in Grossenordnungen beudeuten soll. In erster Näherung findet man die h durch Lösen der Gleichungen

$$\overline{G}(\overline{h}_{sv}) = 0 \ldots \tag{3}$$

Setzt man obige Entwicklung in (2) ein, so erhält man in erster Näherung die Gleichungen

$$L(\overline{\overline{G}}(\overline{h})) = 0 \ldots \tag{4}$$

Dies sind quadratische Gleichungen für die $\overline{h}$, welche im Allgemeinen (beim Nichtbestehen strenger Identitäten) *nicht* Folgerungen aus (3) sind. Diese Gleichungen (4) beschränken nun die Lösungen von (3) so empfindlich, dass nicht darum zu denken ist, dass in der Natur eine derartige Beschränkung zutreffen könnte. Davon habe ich mich durch Beispiele überzeugt.

Es wäre aber doch wünschenswert, wenn Sie hierüber eine allgemein aufklärende Betrachtung anstellen würden.

Noch eine Bemerkung hierüber. Es mögen nun die angedeuteten strengen Identitäten existieren, das System also in Involution sein.

Es gelten strenge die Gleichungen (2). Da nun im Falle meines Gleichungssystems die $L(\ )$ einfache Divergenzen sind, kann man der Gauss'schen Satz anwenden und erhält einen Oberflächensatz

$$\int \overline{\overline{G}}_n df = 0.$$

Wenn nun *an der Fläche* (nicht aber im Innern) die $h_{sv} - \delta_{sv}$ klein sind, so kann man in diesem die h durch die $\overline{h}$ ersetzen und erhält für die erste Näherung eine nicht triviale Integralbedingung. (Diese liefert im Falle der Riemann-Relativitätstheorie die Bewegungsgleichungen). Was sich bei meiner neuen Theorie folgern lässt, habe ich noch nicht nachgesehen.

Mit den herzlichsten Wünschen für 1930 bin ich

Ihr dankbarer

A. Einstein

98

include no rigourous identities then one can show the following remarkable fact.

One can generate a rigorous solution of the equations in the form

$$h_{sv} = \delta_{sv} + \bar{h}_{sv} + \bar{\bar{h}}_{sv} + ...,$$

which stands for an expansion in orders of magnitude. In the first approximation h can be found by solving the equations

$$\overline{G}(\bar{h}_{sv}) = 0. \tag{3}$$

Substituting the series solution above in (2), one obtains, in the first approximation, the equations

$$L(\overline{\overline{G}}(\bar{h})) = 0. \tag{4}$$

These are quadratic equations for $\bar{h}$ which, in general, (where there are no rigorous identities) are *not* consequences of (3). Equations (4) now restrict the solutions of (3) so severely that it is impossible to believe that Nature would have hit upon such a drastic restriction. I have convinced myself of this by examples.

But it would still be desirable if you were to draw up a general clarifying study.

One more remark on this subject. Were it to be arranged that the above mentioned rigorous identities do exist, then the system would be in involution.

Equations (2) hold rigorously. Since in the case of my system of equations the L( ) are simple divergences, one can apply Gauss's Theorem and obtain a surface integral relation

$$\int \overline{\overline{G}}_n df = 0.$$

If now *on the surface* (but not inside it) $h_{sv} - \delta_{sv}$ is small, then one can replace h by $\bar{h}$ and obtain, in the first approximation, a non-trivial integral relation. (This supplies the equations of motion in the case of the Riemannian relativity theory.) I have not yet looked at what follows from my new theory.

With best wishes for 1930,
Gratefully yours,

A. Einstein

XV

Le Chesnay (S. et O.)
27 avenue de Montespan,
le 3 janvier 1930

Cher et illustre Maître,

J'ai bien reçu vos deux lettres; je suis tout à fait heureux que mon manuscrit vous ait intéressé et que vous estimiez ma théorie susceptible de rendre service. Elle est relativement peu connue, probablement parce que je l'ai publiée sous la forme qui se rapporte aux systèmes d'équations aux différentielles totales; certains mathématiciens savent que j'en ai tiré des résultats importants, par exemple la théorie de la structure des groupes continus infinis. Mais sous la forme que je vous ai soumise dans ma note, elle atteindrait évidemment un plus grand public et je vais m'occuper d'en publier l'essentiel.

Je vous remercie des réponses aux questions que je m'étais permis de vous poser, d'abord sur l'existence physique des solutions locales des systèmes différentiels, ensuite sur le problème de la substance. J'aurais désiré vous demander encore comment vous expliquez le fait que, dans la première approximation de votre système, vous ne trouvez que les équations du champ dans le *vide*: c'est sans doute que la matière et l'électricité sont dues à des condensations de champ qui échappent, pour cela même, à la première approximation?

Je vais maintenant tâcher de répondre aux questions que vous m'avez posées.

1) Je ne crois pas qu'il y ait des systèmes en involution qui soient plus déterminés que le vôtre. A priori il pourrait arriver que toute solution à 4 dimensions soit complètement déterminée par une solution à 2 dimensions, ou même à 1 dimension, ou même à 0: mais il me semble — toujours sous la réserve que je n'ai fait qu'une étude sommaire et non définitive — que ces différents cas ne peuvent se présenter, si le système est de la forme analytique que nous avons admise a priori: équations linéaires par rapport aux dérivées premières des $\Lambda_{\alpha\beta}^{\gamma}$ (et système en involution).

A XV

Le Chesnay (S. et O.)
27 avenue de Montespan,
3 January 1930

Cher et illustre Maître,

I have received both your letters. I am very happy that my manuscript has interested you and that you think my theory is likely to be of some help. It is relatively little known, probably because I published it in a form that refers to systems of total differentials. Some mathematicians know that I have derived important results from it, for example the theory of the structure of infinite continuous groups. But in the form that I submitted to you in my note it would, of course, reach a wider public, and I shall publish the essentials of it.

I thank you for the answers to the questions I raised, on the physical existence of the local solutions of the differential systems, and on the problem of a substance. I would have also liked to ask you how you explain the fact that, in the first approximation to your system, you find only the vacuum field equations; it's probably because matter and electricity are due to field condensations which for this very reason, evade the first approximation.

I shall now try to answer your questions.

1) I don't think that there are systems in involution that are more determined than yours. *A priori*, it could happen that any 4-dimensional solution would be completely determined by a 2-dimensional solution, or even a 1-dimensional one, or even a 0-dimensional one: but it seems to me — though I have made only a hasty and non-definitive investigation — that these different cases cannot arise if the system has the analytical form that we have admitted *a priori*: linear equations with respect to the first derivatives of the $\Lambda_{\alpha\beta}^{\gamma}$ (and the system in involution).

2) As far as the *singularity-free* solutions are concerned, it seems to me, the question is extremely difficult. In fact, there are two questions, to all appearence independent but actually intimately

2) En ce qui concerne les solutions *sans singularité*, la question est extrêmement difficile, me semble-t-il. En réalité, il y a deux questions en apparence indépendantes, mais en réalité intimement liées l'une à l'autre. 1° Quel est, au point de vue de l'Analysis situs, l'espace, ou le continuum, dans lequel nous voulons localiser les phénomènes? 2° ce continuum étant choisi, quelles sont les solutions sans singularité dans ce continuum? Il est très possible que l'existence de solutions sans singularité impose des conditions purement topologiques au continuum, exige par exemple que ce continuum soit fermé, comme l'espace sphérique à 4 dimensions.

On a un aperçu de ces difficultés quand on considère le système

$$\Lambda^{\gamma}_{\alpha\beta;\mu} = 0 \tag{1}$$

qui donne naissance aux espaces représentatifs des transformations d'un groupe fini et continu. La résolution de ce système revient d'abord à la recherche d'un système de constantes Λ^k_{ij} satisfaisant aux relations

$$\Lambda^m_{ij}\Lambda^h_{km} + \Lambda^m_{jk}\Lambda^h_{im} + \Lambda^m_{ki}\Lambda^h_{jm} = 0 \tag{2}$$

(équations de structure de S. Lie). Supposons trouvé un système de constantes satisfaisant aux relations (2); les constantes représentent les composantes de la torsion *rapportées à un n-Bein déterminé.* Il n'est pas évident a priori qu'il existe un espace à *n* dimensions dans lequel on puisse trouver des fonction h_{sv} partout régulières conduisant à la torsion considérée; ce théorème est cependant exact et j'en donne une démonstration résumée dans un prochain fascicule du *Mémorial des Sciences mathématiques* [10]; mais cette démonstration repose sur des théorèmes faisant intervenir toute la théorie de la structure des groupes. L'espace dans lequel le groupe existe dépend donc, *au point de vue topologique*, des constantes Λ^k_{ij} et chaque choix de ces constantes donne un espace (ou une famille d'espaces) topologiquement défini. En définitive donc, *chaque solution privée de singularité du système* (1) *crée, au point de vue topologique, le continuum dans lequel elle existe*, et ces solutions sans singularités correspondent aux solutions algébriques des relations (2).

Je me suis étendu sur ce cas particulier, parce qu'il donne probablement un avant-goût de la difficulté du problème général. En tous

related. 1) What, from the point of view of Analysis Situs, is the space or the continuum in which we want to localize the phenomenona? 2) This continuum being chosen, what are the singularity-free solutions in this continuum? It is quite possible that the existence of singularity-free solutions imposes purely topological conditions on the continuum. They may require, for instance, that this continuum be closed, as in a 4-dimensional spherical space.

One gets an idea of these difficulties when considering the system

$$\Lambda^{\gamma}_{\alpha\beta;\mu} = 0 \tag{1}$$

that gives rise to the spaces representing the transformations of a finite continuous group. The resolution of this system amounts to first finding a system of constant Λ^{k}_{ij} which satisfy the relations

$$\Lambda^{m}_{ij}\Lambda^{h}_{km} + \Lambda^{m}_{jk}\Lambda^{h}_{im} + \Lambda^{m}_{ki}\Lambda^{h}_{jm} = 0 \tag{2}$$

(structure equations of S. Lie). Let us assume that a system satisfying relation (2) has been found; then the constants represent the components of the torsion *in a given n-Bein*. A priori, it is not obvious that there exists an *n*-dimensional space in which one can find throughout regular functions $h_{\alpha v}$ leading to the given torsion; this theorem is nevertheless true, and I give an outline proof of it in an upcoming issue of the *Mémorial des Sciences Mathématiques* [10]; but the proof relies on theorems involving the whole theory of the structure of groups. The space in which the group exists, therefore depends, *from the topological point of view*, on the constants Λ^{k}_{ij}, and every choice of these constants gives a space (or a family of spaces) which is topologically defined. In short, *every singularity-free solution of system* (1), *creates from the topological point of view*, the continuum in which it *exists*; and these singularity-free solutions correspond to the algebraic solutions of relations (2). I have dwelt on this particular case, because it give a hint of the difficulty of the general problem. In any case, I know of no method nor even any sketch of a method, permitting us to deal with this problem.

As far as your system is concerned, group theory gives a certain number of singularity-free solutions, but these solutions are *isolated*, they exist in an open space, e.g. Euclidean space. One obtains them by

les cas je ne connais aucune méthode, ni même aucun embryon de méthode, permettant d'aborder ce problème.

En ce qui concerne votre système (la théorie des groupes permet de trouver un certain nombre de solutions sans singularité; mais ces solutions sont *isolées*), elles existent dans un espace ouvert, comme l'espace euclidien. On les obtient en cherchant les systèmes de constantes Λ_{ij}^{k} satisfaisant aux relations (2) et aux relations

$$\Lambda_{ik}^{m} \, \Lambda_{km}^{j} = 0 \, ;$$

je pourrais vous en indiquer quelques-unes si cela vous intéresse; mais je doute que vous puissiez en tirer un parti quelconque et les interpréter physiquement. Chacune de ces solutions correspond à un groupe à 4 paramètres; malheureusement aucune ne correspond au groupe le plus intéressant lié à l'espace sphérique à 3 dimensions et au temps indéfini.

3) J'arrive à la question relative aux systèmes qui ne sont pas en involution. L'exemple que vous me donnez

$$G_{\mu}^{\alpha} \equiv \Lambda_{\mu\underline{\nu};\nu}^{\alpha} = 0$$

est malheureusement mal choisi, car ces 16 équations forment un système en involution à cause des 4 identités

$$G^{\mu\alpha}{}_{;\mu} + \Lambda_{\rho\sigma}^{\alpha} G^{\rho\sigma} \equiv 0;$$

son indice de généralité est 16 [1].

Si nous prenons un système qui ne soit pas en involution, il existe un théorème fondamental, c'est le suivant. Si l'on prolonge le système un certain nombre de fois par l'adjonction, comme nouvelles fonctions inconnues, des dérivées jusqu'à un certain ordre des fonctions données, *on finit par arriver*

soit à un système incompatible
soit à un système en involution.

On peut donc *dans tous les cas* savoir quel est le degré de généralité de la solution du système.

---

1. Ici E. Cartan perd de vue les termes quadratiques: $G_{\mu}^{\alpha} = \Lambda_{\mu\underline{\nu};\nu}^{\alpha} + \Lambda_{\mu\underline{\tau}}^{\sigma} \Lambda_{\sigma\tau}^{\alpha}$; voir lettre XVI.

looking for the systems of constants Λ_{ij}^k satisfying relations (2) and the relations

$$\Lambda_{ik}^m \, \Lambda_{km}^j = 0 \, ;$$

I could show you some if you are interested, but I doubt you will be able to make any use of them or interpret them for physics. Each of these solutions corresponds to a 4-parameter group, but unfortunately none correspond to the most interesting group, that linked to a 3-dimensional spherical space and indefinite time.

3) I consider now the question concerning systems that are not in involution. Unfortunately the example you give me,

$$G_\mu^\alpha \equiv \Lambda_{\mu\nu;\nu}^\alpha = 0$$

is badly chosen, as these 16 equations form a system in involution because of the 4 identities

$$G^{\mu\alpha}{}_{;\mu} + \Lambda_{\rho\sigma}^\alpha G^{\rho\sigma} \equiv 0;$$

Its generality index is 16 [1].

If we take a system that is not in involution, there exists the following fundamental theorem. If one extends the system a certain number of times by adding, as new unknown functions, the derivatives up to a certain order of the given functions, *one finally obtains*

either an incompatible system
or a system in involution.

Hence, *in all cases*, we can determine the generality index of the solution of the system.

Now, it may happen that a system allows singular solutions, this doesn't mean that the solutions have analytical singularities but it does mean that all the sections (the 3-dimensional ones, for instance) are characteristics. Your system has no such singular solution, but in geometry such singular solutions often occur, and it may happen that they are more important than the general solutions.

There remains the unsettled question of finding all systems in involution having the same degree of determination as yours. Perhaps

Il peut du reste arriver qu'un système admette des solutions singulières; cela ne veut pas dire des solutions admettant des singularités analytiques, mais cela signifie que toutes les sections (à 3 dimensions par exemple) sont caractéristiques. Votre système n'admet aucune solution singulière de cette nature; mais en géométrie de pareilles solutions singulières se présentent fréquemment et il peut arriver qu'elles soient plus importantes que les solutions générales.

Reste la question non tranchée de la recherche de tous les systèmes en involution ayant le même degré de détermination que le vôtre. Peut-être existe-t-il une suite continue de systèmes de cette nature, suite contenant votre système et le système aux R_{ik}?! J'espère pouvoir me remettre à ce problème, qui exige malheureusement beaucoup de patience et aussi une absence complète de fautes de calcul! Après beaucoup de calculs, on n'a plus qu'à identifier à 0 une forme cubique des $\Lambda_{\alpha\beta}^{\gamma}$ qui ne tient pas plus d'une page! Mais il y a peut-être une méthode évitant tous ces calculs; en tous cas je l'ignore.

Veuillez agréer, cher et illustre Maître, l'expression de mes sentiments les plus dévoués.

E. Cartan

there exists a continuous sequence of systems of that kind, a sequence containing your system and the system with R_{ik}?! I hope to be able to come back to this problem that unfortunately, requires much patience as well as a complete absence of mistakes! After many calculations, there remains only to set to zero a form cubic in the $\Lambda^{\gamma}_{\alpha\beta}$, and that doesn't take more than a page! Perhaps there is a way to avoid all these calculations; but if there is, I don't know it...

E. Cartan

XVI

7-I-30

Verehrter Herr Cartan!

Ich freue mich, dass Sie Ihre Theorie der Involutions-Systeme publizieren wollen. Aber thun Sie es bitte mit derselben Ausführlichkeit wie mir gegenüber. Mir ist das Studium auch *so* gar nicht so leicht geworden. Und es ist doch wirklich eine Theorie von fundamentaler Beudeutung.

Die Frage wegen der Nicht-Linearität der Gleichungen bei Berücksichtigung von Elektrischen und Massen-Dichten haben Sie in Ihrem Brief schon in meinem Sinne beantwortet. Begründung: bei Intervention von Massendichten gilt das Superpositions-Prinzip nicht; also Nicht-Linearität der entsprechenden Gleichungen.

Im Übrigen muss ich mich nachträglich für die Rücksichtslosigkeit meines Fragens entschuldigen. Es gilt da das schöne Sprichwort: Ein Narr kann mehr fragen als ein Kluger beantworten.

Es freut mich, dass Sie eine gewisse Sicherheit fühlen in der Antwort, dass Involutionssysteme von geringerer Freiheit nicht existieren. Ich habe nämlich auch angestrengt und vergeblich danachgesucht. Natürlich beschränkte ich mich dabei auch auf Systeme, die in den zweiten Ableitungen linear sind.

Über die Zusammenhangs-Verhältnisse des Raumes kann ich nichts aussagen, wohl aber scheint es unvermeidlich, Singularitätsfreiheit der Lösungen zu verlangen. Abgesehen davon, dass dies als die höchste Forderung an sich wünschbar ist, besteht auch eine Notwendigkeit speziellerer Art.

Meine Gleichungen sind strenge erfüllt durch den Ansatz für die h_{sv}

$$\begin{matrix} 1 & 0 & 0 & 0 \\ 0 & 1 & 0 & 0 \\ 0 & 0 & 1 & 0 \\ 0 & 0 & 0 & h_{44} \end{matrix} \qquad h_{44} = \frac{1}{1 + \Sigma \frac{\alpha}{r}}$$

A XVI

7.1.30

Dear M. Cartan,

I am delighted that you intend to publish your theory of systems in involution. But please do it in the same detail as you did for me. Learning has become not at *all* easy for me anymore; and this is really a theory of fundamental importance.

In your letter you have already answered the question about the nonlinearity of the equations with respect to electrical and mass densities just as I would have. Argument: because of the appearance of mass densities the superposition principle does not hold; hence nonlinearity of the corresponding equations.

In addition, I must further apologize for the thoughtlessness of my questions. There's a lovely old saying: A fool can ask more questions than a wise man can answer.

I am happy that you feel confidence in your answer that there exists no involutive system of less freedom. I myself have tried in vain to find one. Naturally, I restricted myself to systems which are linear in their second derivatives.

Concerning the connectivity properties of space, I can say nothing, but probably it is unavoidable to demand that the solutions be singularity-free. Apart from the fact that this is a highly desirable requirement in itself, it is also a necessity of a more special kind.

My equations are rigorously satisfied by an h_{sv} of the form

$$
\begin{matrix}
1 & 0 & 0 & 0 \\
0 & 1 & 0 & 0 \\
0 & 0 & 1 & 0 \\
0 & 0 & 0 & h_{44}
\end{matrix}
\qquad
h_{44} = \frac{1}{1 + \Sigma \frac{\alpha}{r}}
$$

Die Summe ist über eine beliebige diskrete Zahl von singulären Punkten zu erstrecken; r ist der Abstand des Aufpunktes von einem dieser singulären Punkte [1].

Würde man solche Singularitäten ($h_{44} = 0$) zulassen, so gäbe es 1) ungeladene Massenpunkte, 2) im Gleichgewicht miteinander stehend, was beides der Erfahrung widerspricht. Die Theorie könnte also nur dann aufrecht erhalten werden, wenn diese beiden Widersprüche mit der Erfahrung beim Ausschluss von Singularitäten verschwinden. Leider scheinen die Korpuskeln in der Natur nicht zentralsymmetrisch zu sein (magnetisches Moment), sodass die Entscheidung der Frage, ob solche singularitätsfreie Lösungen existieren, eine mathematisch recht schwierige ist.

Und dabei ist es klar, dass *vor* der Lösung dieses Problem das Verhalten eines Korpuskels in einem Felde (Bewegungsgesetz) nicht in Angriff genommen werden kann. Und ohne die Lösung *dieses* Problem kann das Zutreffen der Theorie nicht geprüft werden!

Das Schlimme ist, dass unsere theoretischen Physiker nicht mitarbeiten wollen, sondern mich beschimpfen, weil sie für die Natürlichkeit des eingeschlagenen Weges kein Organ haben — (ausser Langevin!).

Als ich Ihre Bemerkung las, dass das System

$$G_\mu{}^\alpha \equiv \Lambda^\alpha_{\mu\underline{\nu};\nu} = 0 \qquad\qquad (I)$$

in Involution sei, bekam ich einen grossen Schrecken. Denn die Nichtexistenz einer Vierer-Identität in diesem Falle bildete eine der Konstatierungen, von denen ich ausgegangen war. Sie müssen sich aber hierin getäuscht haben, denn die Identität

$$G^{\mu\alpha}{}_{;\mu} + \Lambda^\alpha_{\rho\sigma} G^{\rho\sigma} \equiv 0$$

gilt ja für die Grössen

$$G_\mu{}^\alpha \equiv \Lambda^\alpha_{\mu\underline{\nu};\nu} - \Lambda^\sigma_{\mu\underline{\tau}}\Lambda^\alpha_{\sigma\tau} \qquad\qquad (II)$$

---

1. Dans [23] Einstein et Mayer donnent des solutions statiques exactes des équations de champ, en particulier une solution purement gravitationnelle qui est la suivante: si h_s^α, h_4^α sont quatre vecteurs tangents constituant un repère on a

$$h_s^\alpha = \delta_s^\alpha \qquad h_s^4 = 0 \qquad s,a = 1,2,3,$$
$$h_4^\alpha = 0 \qquad h_4^4 = \delta_4^\alpha \sigma$$

avec $\sigma = 1 + \sum_j \dfrac{m_j}{r_j}$, m_j constant; les r_j sont les distances spatiales à l'origine.

The sum ranges over an arbitrary discrete number of singular points, r is the distance of the origin from one of these singular points [1].

Were one to allow such singularities ($h_{44} = 0$), these would represent 1) uncharged point masses 2) at rest with respect to each other, both of which contradict experience. The theory could then only be maintained if these two contradictions with experience were to vanish with the exclusion of singularities. Unfortunately, in Nature corpuscles do not appear to be spherically symmetric (magnetic moment), so to decide the question of whether such singularity-free solutions exist is mathematically quite difficult.

Thus it is clear that *prior* to the solution of this problem the behaviour of a corpuscle in a field (its equations of motion) cannot be tackled. And without the solution of *this* problem, the truth of the theory cannot be tested!

The worst is that our theoretical physicists do not wish to collaborate, but rather abuse me because they have no feeling for the naturalness of this approach (except for Langevin!).

When I read your remark that the system

$$G_\mu{}^\alpha \equiv \Lambda_{\mu\nu;\nu}^\alpha = 0 \tag{I}$$

is in involution, I experienced a great shock. For the non-existence of a 4-identity in this case formed one of the starting-points from which I set out. But you must be mystaken here, for the identity

$$G^{\mu\alpha}{}_{;\mu} + \Lambda_{\rho\sigma}^\alpha G^{\rho\sigma} \equiv 0$$

holds for the quantities

$$G_\mu{}^\alpha \equiv \Lambda_{\mu\nu;\nu}^\alpha - \Lambda_{\mu\tau}^\sigma \Lambda_{\sigma\tau}^\alpha \tag{II}$$

(as you yourself have pointed out) and not for the quantities (I). In fact, in the case of (I), the commutation rule for differentiation yields

$$G^{\mu\alpha}{}_{;\mu} \equiv -\frac{1}{2}\Lambda_{\mu\nu;\sigma}^\alpha \Lambda_{\nu\mu}^\sigma,$$

Les auteurs notent immédiatement la difficulté d'une telle solution: il existe donc une solution exacte où deux ou plusieurs masses, électriquement neutres et à distances arbitraires les uns des autres se trouvent au repos. Ils insistent sur la nécessité de trouver des solutions sans singularité et sur celle de déduire les équations du mouvement des particules à partir des équations de champ.

(wie Sie selbst mich aufmerksam gemacht haben) und nicht für die Grössen (I). In der That liefert (für I) der Vertauschungs-Satz der Differentiationen

$$G^{\mu\alpha}{}_{;\mu} \equiv -\frac{1}{2} \Lambda^{\alpha}_{\underline{\mu\nu};\sigma} \Lambda^{\sigma}_{\nu\mu},$$

und es kann in das Glied der rechten Seite auf keine Weise der Faktor G hinein gebracht werden, wofür ja nur die zyklische fundamentale Identität zur Verfügung stünde.

Die Frage, ob (I) nur gänzlich triviale Lösungen zulasse, erscheint demnach berechtigt. Die primitive Art, wie ich sie nur für meine Zwecke beantwortete, habe ich Ihnen ja mitgeteilt. Haben Sie meine diesbezügliche Argumentation verstanden?

Ihre Behauptung der Zurückfahrbarkeit von Nicht-Involutions-Systemen auf Involutions-Systeme oder widerspruchsvoller Systeme entspricht meiner Vermutung, die ich mit meinem Freunde vergeblich zu bewiesen versuchte. Es wäre sehr gut, wenn Sie diesen Beweis Ihrer Abhandlung anfügen wollten. Denn dies ist ebenfalls ein prinzipiell wichtiger Punkt. Ich bin Ihnen wirklich von Herzen dankbar für die Klarheit, die Sie mir gegeben haben.

Zur Frage der Existenz anderer Gleichungs-Systeme von gleichem Determinationsgrad wie die beiden bekannten glaube und — hoffe ich: nein. Bezüglich meiner Anfrage erinnere ich reuevoll an das obige Sprichwort!

Herzlich grüsst Sie

Ihr

A. Einstein

112

and the factor G can in no way be taken into the term on the right hand side, for which purpose only the cyclic fundamental identity is at our disposal.

The question whether (I) possesses only completely trivial solutions thus appears justified. The crude way in which I answered it — just for my own needs — I have already communicated to you. Did you understand my arguments about this?

Your assertion that non-involutive systems can be reduced either to involutive or to contradictory systems corresponds to my conjecture which my friend and I tried in vain to prove. It would be very good if you were to add this proof to your article, for this, too, is a crucially important point. I am truly thankful to you for the clarity you have given me.

On the question of the existence of another system of equations with the same degree of determination as the two known ones, I believe — and I hope — the answer to be No. As for my questions, I refer back to the old maxim mentioned above.

Kind regards

Yours,

A. Einstein

XVII

8-I-30

Verehrter Herr Cartan!

Ich schäme mich, dass ich Sie schon wieder störe, aber ich will kurz sein und hoffe, dass die hier zu berührende Frage Sie ebenso eletrisieren wird wie mich. Es handelt sich um eine Paradoxie, hinter der etwas Wichtiges verborgen zu sein scheint, betreffend Ihren *Index de généralité* ($\mathscr{I}$) [1].

Es zeigt sich, das $\mathscr{I}$ für die approximierten Gleichungen grösser ist als für die strengen, sowohl bei der alten Theorie der Gravitation als auch bei der neuen Feldtheorie

a) Alte Theorie

streng $\qquad$ $R_{ik} = 0$ $\qquad$ $\mathscr{I} = 8$ (bei festgelegter Koordinaten-wahl)

approximativ $\left. \begin{array}{l} \gamma_{ik,\alpha\alpha} = 0 \\ \gamma_{i\alpha,\alpha} = 0 \end{array} \right\}$ $\mathscr{I} = 12$

$$\left(\gamma_{ik} = \bar{g}_{ik} - \frac{1}{2} \delta_{ik} \bar{g}_{\alpha\alpha}; \; \bar{g}_{ik} = g_{ik} - \delta_{ik} \infty \text{ klein} \right)$$

b) Neue Theorie

streng $\qquad$ $\left. \begin{array}{l} G^{ik} = 0 \\ F^{ik} = 0 \end{array} \right\}$ $\mathscr{I} = 10$ (bei festgelegter Koordinatenwahl)

approximativ $\left. \begin{array}{l} \bar{h}_{ik,\alpha\alpha} = 0 \\ \bar{h}_{i\alpha,\alpha} = 0 \\ \bar{h}_{\alpha k,\alpha} = 0 \end{array} \right\}$ $\mathscr{I} = 17$

Es ist also klar, dass nicht zu jeder Lösung der approximativen Glei-chungen eine strenge Lösung gehört. Es muss alzo zu den approxima-

---

1. Voir note 4, Lettre VII.

114

A XVII

8-1-30

Dear M. Cartan,

I am ashamed to bother you again so soon, but I will be short and I hope the question raised here will electrify you as much as it has done me. It concerns a paradox, behind which there seems to lie something important touching your *indice de généralité* ($\mathscr{I}$) [1].

It can be shown that $\mathscr{I}$ is bigger for the approximate equations than for the full equations, both in the case of the old theory of gravitation and the new theory,

a) Old theory

full equations $R_{ik} = O$ $\mathscr{I} = 8$ (with fixed choice of coordinates)

approximate equations
$$\left.\begin{array}{l} \gamma_{ik,\alpha\alpha} = 0 \\ \gamma_{i\alpha,\alpha} = 0 \end{array}\right\} \ \mathscr{I} = 12$$

$$\left(\gamma_{ik} = \bar{g}_{ik} - \frac{1}{2}\delta_{ik}\bar{g}_{\alpha\alpha}; \ \bar{g}_{ik} = g_{ik} - \delta_{ik} \infty^{ly} \text{ small}\right)$$

b) New theory

full equations
$$\left.\begin{array}{l} G^{ik} = 0 \\ F^{ik} = 0 \end{array}\right\} \begin{array}{l} \mathscr{I} = 10 \text{ (with fixed choice of} \\ \text{coordinates)} \end{array}$$

approximate equations
$$\left.\begin{array}{l} \bar{h}_{ik,\alpha\alpha} = 0 \\ \bar{h}_{i\alpha,\alpha} = 0 \\ \bar{h}_{\alpha k,\alpha} = 0 \end{array}\right\} \ \mathscr{I} = 17$$

Thus, clearly, it is not true that to every solution of the approximate equations there belongs a rigorous solution. Hence, supplementary

---

Signalons ici que $\mathscr{I} = 4$ dans le cas des équations R_{ik} exactes ou approchées, et $\mathscr{I} = 12$ dans la théorie d'Einstein du parallélisme absolu.

tiven Gleichungen supplementäre Bedingungen geben, die aussagen, dass die Lösung einer strengen Lösung benachbart ist. Es ist klar, dass die Aufstellung dieser Zusatzbedingungen die Übersicht über den Gegenstand enorm fördern würde. Diesen Faden müssen wir gut festhalten!

Herzlich grüsst Sie

Ihr

A. Einstein

conditions must be added to the approximate equations to insure that the solution has a neighbouring rigorous solution. It is clear that the formulation of these additional conditions would enormously advance the understanding of the subject. We must hold tight to this thread!

Kind regards.

Yours,

A. Einstein

XVIII

10-I-30

Verehrter Herr Cartan!

Ich glaube, Ihnen zwei Punkte mitteilen zu müssen, wo wir — wie mir scheint — verschiedener Meinung sind.

1) Die frühere Relativitätstheorie mit Berücksichtigung des elektromagnetischen Feldes, welche so geschrieben werden kann

$$R_{ik} - \frac{1}{2} g_{ik} R = -T_{ik},$$

$$\frac{\partial f_{ik}}{\partial x^l} + \frac{\partial f_{kl}}{\partial x^i} + \dots = 0,$$

$$\frac{\partial f^{ik} \sqrt{-g}}{\partial x^k} = 0,$$

hat ebenfalls $\mathscr{I} = 12$. Die Gleichung $R_{ik} = 0$ hat nämlich $\mathscr{I} = 8$. Dazu kommen dann wegen der zwei Divergenzgleichungen des elektrischen und magnetischen Feldes noch die Angabe von $6 - 2$ Komponenten (z.B. n_1, n_2, f_1, f_2) für den Schnitt $x^4 = $ konst. Dann ist die Fortsetzung völlig bestimmt.

Ich möchte Ihnen kurz die Gründe sagen, warum die neuen Gleichungen aus formalen Gründen mit besser gefallen. Erstens ist die Feldauffassung hier keine einheitliche, indem die g und ϕ nichts miteinander logisch zu thun haben. Zweitens haben linke und rechte Seite der Gravitations-Fedgleichungen logisch nichts miteinander zu schaffen. Drittens kann diese Theorie niemals elektrische Massen anders auffassen wie als Singularitäten — das heisst überhaupt nicht auffassen — und zwar wegen der letzten Gleichung des dritten Systems, welche reinen Divergenz-Charakter auf der linken Seite hat. Alle diese Übelstände fallen bei der neuen Theorie weg.

118

A XVIII

10-I-30

Dear M. Cartan,

I believe I ought to communicate two points to you where — it seems to me — we are of different minds.

1) The earlier theory of relativity, including the electromagnetic field, which can be written

$$R_{ik} - \frac{1}{2} g_{ik} R = -T_{ik},$$

$$\frac{\partial f_{ik}}{\partial x^l} + \frac{\partial f_{kl}}{\partial x^i} + \ldots = 0,$$

$$\frac{\partial f^{ik} \sqrt{-g}}{\partial x^k} = 0,$$

likewise has $\mathscr{I} = 12$. The equations $R_{ik} = 0$ have $\mathscr{I} = 8$. It then follows that, with the two divergence equations for the electric and the magnetic field, one can still specify $6 - 2$ components (e.g. n_1, n_2, f_1, f_2) on the cross-section $x^4 = const.$ Then the continuation is fully determined.

I would like to tell you briefly the reasons why, on formal grounds, the new equations please me more. First, the field interpretation is not unified in that g and ϕ have logically nothing to do with each other. Second, the left and right hand sides of the gravitational field equations have no logical connection. Third, this theory can never express charged masses otherwise than as singularities — that is, they cannot express them at all — and this because of the last equation of the third set, whose left hand side has the form of a pure divergence. All these drawbacks are overcome in the new theory.

2) Die von Ihnen ins Auge gefassten Gleichungen

$$R_{ik} = 0,$$

$$\phi_\alpha = \frac{\partial \log \psi}{\partial x^\alpha},$$

$$S_\alpha = \frac{1}{\psi} \frac{\partial \chi}{\partial x^\alpha},$$

haben *nicht* $\mathscr{I} = 12$, sondern $\mathscr{I} = 16$. Die Gleichungen R_{ik} für sich allein haben nämlich $\mathscr{I} = 8$, wenn man die g_{ik} allein als Feldvariable ansieht. Man kann nun die h_{sv} aus den g_{ik} und 6 weiteren Variablen $\lambda_{(\mu)}$ ausgedrückt denken. Dann werden die zweite und dritte Feldgleichung linear in den ersten Ableitungen nach x_4 von $\lambda_{(\mu)}$, ψ und χ. Alles ist also bestimmt, wenn neben den auch bei reiner Gravitation zu gebenden 8 Funktionen $g_{\mu v}$, bezw. $g_{\mu v,\alpha}$ für den betrachteten Schnitt $x^4 = $ konst., noch jene 8 Funktionen $\lambda_{(\mu)}$, ψ, χ gegeben werden. Es ist also $\mathscr{I} = 8 + 8$ und nicht 12.

Es ist also bis jetzt überhaupt kein zweites Gleichungssystem von *Ind. d. généralité* $\mathscr{I} = 12$ aufgezeigt.

Nehmen Sir mie meinen oberflächlichen Brief von neulich nicht übel sondern haben Sie auch in Zukunft freundlich Geduld mit Ihrem

A. Einstein

P.S. Es ist merkwürdig, dass die *Mathematische Annalen* eine so schreckliche Verstopfung haben, dass sie in so vielen Monaten nicht ausscheiden, was sie zu sich genommen haben.

2) The equations which you have in mind;

$$R_{ik} = 0,$$

$$\phi_\alpha = \frac{\partial \log \psi}{\partial x^\alpha},$$

$$S_\alpha = \frac{1}{\psi}\frac{\partial \chi}{\partial x^\alpha},$$

do *not* have $\mathscr{I} = 12$, but rather $\mathscr{I} = 16$. The equations R_{ik} alone have $\mathscr{I} = 8$, if one views the g_{ik} alone as field variables. Now, one can express the $h_{s\nu}$ from the g_{ik} and six further variables $\lambda_{(\mu)}$. Then the second and third field equations will be linear in the first derivatives of $a_{(\mu)}$, ψ and χ with respect to x_4. Then everything is fixed when, besides the 8 functions $g_{\mu\nu}$, resp. $g_{\mu\nu,\alpha}$ — which also must be given in the pure gravitational case — on the chosen section $x^4 = const.$, there are also given another 8 functions, $\lambda_{(\mu)}$, ψ, χ. Hence $\mathscr{I} = 8 + 8$ and not 12.

Thus, up to now, no second system of equations has presented itself with an *ind. d. généralité* $\mathscr{I} = 12$.

Do not take my perfunctory letter of the other day badly, but for the future be kind enough to be patient with your

A. Einstein

P.S. It is remarkable that the *Mathematische Annalen* has such terrible constipation that, after so many months, it has not been able to excrete what it has absorbed.

XIX

11-I-30

Une carte postale

Prof. Dr. Cartan
27, avenue Montespan
Le Chesnay (près de Paris)
Frankreich

Pater Peccavi! [1]

    Man kann die schönste Theorie verkehrt anwenden! Verzeihung

Herzlichen Gruss

A. E.

(Das eine mal habe ich das Koordinatensystem durch Differentialgleichungen, das andere mal durch Festlegen von 4 Variabeln bestimmt.)

---

1. Pater peccavi, soit: mon Père, j'ai péché. C'est la formule de la confession dans la religion catholique.

Absender: ....

Wohnort: ....
Straße, hausnummer,
Gebäudeteil, Stockwerk

Postkarte

Très pressé!

Man kann die schöne
Theorie verkehrt anwenden!

Verzeihung

Herzlichen Gruß
H. B.

(An und ab habe ich das
verschwindungen durch Differential
gleichungen, besonders Maßsysteme

Prof. Dr. Cartan

27 avenue Montespan

in Le Chesnay (près de Paris)

Frankreich.

Straße, hausnummer,
Gebäudeteil, Stockwerk

⊕ (9.26)   C 154   Din 476

11 JANUARY 1930

A XIX

[Post card — postmarked 11-1-30]

Prof. Dr. Cartan
27 avenue Montespan
Le Chesnay (*près de Paris*)
France

Pater peccavi! [1]

Even the most beautiful theory can be applied wrongheadedly!
Forgive me.

Kind regards.

A. E.

(One time I fixed the coordinate system by means of differential equations, the other time by fixing 4 variables.)

XX

Le Chesnay (S. et O.)
27 avenue de Montespan,
le 11 janvier 1930

Cher et illustre Maître,

J'ai bien reçu coup sur coup vos deux lettres et votre carte. Je me proposais justement de vous écrire un mot quand cette dernière est arrivée. Est-ce bien à moi qu'il convient de confesser vos péchés? Je puis bien dire à mon tour: *Pater, peccavi*! car j'ai été bien étourdi avec votre question sur le système $\Lambda^{\beta}_{\alpha\mu;\mu} = 0$. Quand j'ai lu votre lettre, je me suis dit: « Voilà une question à examiner ». Puis quand le lendemain j'y ai réfléchi, j'ai totalement oublié que les termes quadratiques ne devaient pas figurer dans les premiers membres des équations, et à vrai dire je n'ai même pas songé à me reporter aux termes de votre lettre! Vous voudrez bien m'excuser de cette étourderie.

En ce qui concerne votre argumentation au sujet de l'impossibilité du système en question, j'ai cru comprendre qu'elle reposait au fond sur l'existence, pour les équations approchées, de solutions qui ne peuvent être les premières approximations de solutions rigoureuses. Si c'est bien votre pensée, cela se rattache à l'observation de votre dernière lettre. C'est qu'en effet il est facile de prouver que *si un système est en involution, toute solution des équations en première approximation peut être regardée comme la première approximation d'une infinité de solutions rigoureuses.* On peut le montrer de différentes manières; la plus facile à comprendre est peut-être la suivante.

Supposons qu'on parte d'une solution particulière, que nous pouvons toujours supposer être $f_\alpha = 0$. Les équations en première approximation, qui donnent les solutions infiniment voisines, s'obtiennent en ne conservant que les termes linéaires en f_α. Remplaçons alors dans les équations f_α par

$$f_\alpha = t f_\alpha^{(1)} + t^2 f_\alpha^{(2)} + t^3 f_\alpha^{(3)} + \ldots,$$

A XX

Le Chesnay (S. et O.)
27 avenue de Montespan,
11 January 1930

Cher et illustre Maître,

I received your two letters and your card one after the other. I was just intending to write you a word when the last one arrived. Is it really to me that you should confess your sins? I too might well say: " *Pater peccavi!* " for I have been quite irresponsible about your question on the system $\Lambda^{\beta}_{\alpha\mu;\mu} = 0$. When I read your letter, I said to myself: that's a question to be investigated. And when, the next day, I thought it over, I had completely forgotten that the quadratic terms need not appear on the left hand sides of the equations; to tell the truth, I did not even dream of referring back to your letter. Will you please forgive my thoughtlessness.

Concerning your arguments about the impossibility of this system: as I understand it, they rely heavily on the existence of solutions for the approximate equations which cannot be first approximations of rigorous solutions. If this is really what you have in mind, it is connected with the remark in your last letter. It is indeed easy to prove that *if a system is in involution, any solution of the first approximation equations can be regarded as the first approximation of infinitely many rigorous solutions.* One can show this in several ways; perhaps the easiest one to understand is the following.

Assume that one starts from a particular solution, one we can always assume to be $f_\alpha = 0$. The first approximation equations giving the infinitely close neighbouring solutions are obtained by keeping only the terms linear in f_α. Replace then in the equations f_α by

$$f_\alpha = tf_\alpha^{(1)} + t^2 f_\alpha^{(2)} + t^3 f_\alpha^{(3)} + ...,$$

125

t étant un paramètre variable, et considérons les systèmes successifs obtenus en prenant les termes en t, les termes en t^2, etc., soient

$$E^{(1)} = 0, \tag{1}$$

$$E^{(2)} = 0, \tag{2}$$

$$\dots\dots$$

ces systèmes. Les équations $E^{(1)} = 0$ ne contiennent que les fonctions inconnues $f_\alpha^{(1)}$; ce sont les équations de la première approximation (*équations aux variations* de Poincaré et Darboux). Les équations $E^{(2)} = 0$ contiennent les fonctions inconnues $f_\alpha^{(2)}$ (et aussi les $f_\alpha^{(1)}$), et ainsi de suite.

Étant donné l'existence des identités qui assurent que le système primitif est en involution, les équations (1) seront en involution; les équations (2), *quand on y remplace les $f_\alpha^{(1)}$ par une solution de* (1), seront aussi en involution; de même les équations (3), quand on y remplace les $f_\alpha^{(1)}$ et $f_\alpha^{(2)}$ par une solution de (1) et (2), et ainsi de suite. Toute solution de (1) peut donc être regardée, et d'une infinité de manières, comme le premier terme du développement par rapport aux puissances de t d'une solution rigoureuse des équations données.

Cette démonstration a l'inconvénient de laisser en suspens les questions de convergence, mais on peut la présenter sous une forme telle que cette objection ne soit plus possible.

J'avais déjà remarqué que vos équations approchées comportent un indice de généralité trop grand, et j'avais été sur le point de vous en parler dans ma dernière lettre, mais comme elle était déjà assez longue, j'avais reporté cette question à plus tard. La manière dont vous choisissez les coordonnées nous laisse encore beaucoup d'arbitraire, puisque vous pouvez remplacer x_i par $x_i + \varepsilon\xi_i$, où les fonctions ξ_i satisfont aux équations

$$\xi_{i,kk} = 0, \quad \xi_{k,k} = 0. \tag{3}$$

Il en résulte que deux champs de gravitation pour lesquels les composantes $g_{\alpha\beta}$ diffèrent de fonctions de la forme $\dfrac{\partial \xi_\alpha}{\partial x_\beta} + \dfrac{\partial \xi_\beta}{\partial x_\alpha}$, où les ξ satisfont aux équations, correspondent néanmoins au *même* espace de Riemann à parallélisme absolu, c'est-à-dire ne doivent pas physiquement différer l'un de l'autre. Il est vrai que si ces deux champs de gravitation s'annulent à l'infini, les fonctions ξ_α sont nécessairement nulles. Il y a néanmoins quelque chose d'un peu déconcertant, c'est

126

where t is a parameter, and consider the successive systems obtained by taking the terms in t, in t^2 and so on... Let

$$E^{(1)} = 0, \tag{1}$$

$$E^{(2)} = 0, \tag{2}$$

$$\cdots\cdots$$

be these systems. The equations $E^{(1)} = 0$ contain only the unknown functions $f_\alpha^{(1)}$; they are the first approximation equations ("*équations aux variations*" of Poincaré and Darboux). The equations $E^{(2)} = 0$ contain the unknown functions $f_\alpha^{(2)}$ (and also the $f_\alpha^{(1)}$) and so on.

As a result of the existence of the identities that ensure the original system is in involution, equations (1) will be in involution; equations (2) will also be in involution *when the $f_\alpha^{(1)}$ are replaced by a solution of* (1). The same is true for the equations (3) when the $f_\alpha^{(1)}, f_\alpha^{(2)}$ are replaced by a solution of (1) and (2), and so on. Any solution of (1) can thus be regarded, in a infinite number of ways, as the first term of a power series expansion with respect to t of a rigorous solution of the given equations.

This proof has the disadvantage of leaving aside the question of convergence, but it can be formulated in such a way that this objection is no more valid.

I had already noticed that your approximate equations yield a generality index which is too large and I was on the verge of talking to you about it in my last letter, but as it was already quite long I postponed the question. The results of your choice of coordinates leaves a great deal of arbitrariness, since one can replace x_i by $x_i + \varepsilon\xi_i$, where the functions ξ_i satisfy the equations

$$\xi_{i,kk} = 0, \quad \xi_{k,k} = 0. \tag{3}$$

Hence, two gravitational fields for which the components $g_{\alpha\beta}$ differ by functions of the form $\dfrac{\partial\xi_\alpha}{\partial x_\beta} + \dfrac{\partial\xi_\beta}{\partial x_\alpha}$, where the ξ satisfy the equations, correspond nevertheless to the *same* Riemannian space with absolute parallelism; that is they should not differ physically from each other. It is true that if these two gravitational fields vanish at infinity, the functions ξ_α are necessarily zero. Nevertheless, there is the frustating fact that it seems impossible to find, in first approximation, a system of normal coordinates which are invariantly and

127

qu'il semble impossible de trouver, en première approximation, un système de coordonnées normales déterminé sans ambiguïté d'une manière invariante, c'est-à-dire par des conditions invariantes par un déplacement quelconque dans l'espace euclidien. Tout cela est encore assez obscur pour moi.

J'ai repris mes calculs pour la détermination des systèmes de 22 équations, mais sans aboutir encore à quelque chose de définitif. Je regarde cependant comme très vraisemblable qu'il ne doit pas y avoir d'autre système que ceux que nous connaissons, ou tout au moins qu'il ne peut y avoir que des systèmes *isolés*.

Veuillez agréer, cher et illustre Maître, l'expression de mes sentiments les plus dévoués.

E. Cartan

unambiguously defined, that is, defined by conditions that are invariant under any displacement in Euclidean space. All this is still rather obscure to me.

I have returned to my calculations for the determination of systems with 22 equations but without yet reaching any conclusion. Nevertheless, I consider it highly probable that there are no systems other than those we know, or, at least, that there can exist only *isolated* systems...

E. Cartan

XXI

Le Chesnay (S. et O.)
27 avenue de Montespan
le 12 janvier 1930

Cher et illustre Maître,

Un petit mot seulement pour vous apprendre que je suis à peu près certain du résultat cherché depuis si longtemps. Il n'y a pas d'autre système déterministe en involution que les deux connus à 22 *équations* (et ceux à 15 et 16 équations que vous rejetez comme pas assez déterminés). Finalement la démonstration peut se présenter sans des calculs trop formidables.

Je n'ai pas encore eu le temps de méditer votre dernière lettre du 10 janvier. Mais je crois que votre conclusion relative à l'indice de généralité du système $R_{ik} = 0$ n'est pas exacte; j'avais calculé il y a très longtemps cet indice et je viens encore de faire le calcul; je trouve $\mathscr{I} = 4$ et non pas 8. Quant à votre système dans lequel entre le tenseur T_{ik}, j'y réfléchirai à loisir et je vous écrirai le résultat de mes réflexions.

Je dois faire après-demain au laboratoire de Langevin une conférence sur l'Introduction géométrique à votre nouvelle théorie; je parlerai surtout de la notion géométrique de parallélisme absolu et de torsion. Elle sera suivie d'une autre où je parlerai peut-être du résultat de mes recherches en collaboration avec vous et sous votre direction, si vous le permettez [1].

Veuillez agréer, cher et illustre Maître, l'expression de mes sentiments les plus dévoués.

E. Cartan

---

1. Outre les publications [12], [13], E. Cartan, dans la « Notice sur les Travaux Scientifiques » [14, p. 111], écrit ceci:

« M. Einstein a indiqué comme susceptible d'être mis à la base de sa théorie unitaire, un système de 22 équations, linéaires par rapport aux dérivées premières des composantes de la torsion et quadratiques par rapport à ces composantes. *Dans une série de trois conférences faites dans l'hiver 1930 au Collège de France* (souligné par nous), j'ai démontré rigoureusement que ce système est compatible et qu'il respecte le déterminisme (au moins mathématique). J'ai plus généralement déterminé par des calculs assez longs, qui font intervenir la théorie de la représentation linéaire des groupes semi-simples et ma théorie des systèmes en involution, tous les systèmes possibles satisfaisant aux conditions admises plus ou moins explicitement par M. Einstein.

A XXI

Le Chesnay (S. et O.)
avenue de Montespan
12 January 1930

Cher et illustre Maître,

Just a word to tell you that I am almost sure of the result I have been searching for such a long time. There are no deterministic systems in involution other than the two known ones with 22 *equations* (and those with 15 and 16 equations that you reject as not deterministic enough). And the proof can be presented without any calculations which are too horrendous.

I have not yet had time to reflect on your last letter of January 10. But I think your conclusion about the generality index of the system $R_{ik} = 0$ is not right; I calculated this index a long time ago, and I have just redone the calculations; I find $\mathscr{I} = 4$ and not 8. Concerning your system which contains the tensor T_{ik}, I'll think it over at leisure and write you the results of my thoughts.

The day after tomorrow, I have to give a lecture on a geometrical introduction to your new theory at Langevin's laboratory; I shall speak especially about the geometrical concepts of torsion and absolute parallelism. It will be followed by another where I shall speak, perhaps, about the results of my research in collaboration with you and under your direction, if you will permit me [1].

E. Cartan

J'ai trouvé ainsi;

1° Des systèmes assez généraux de 15 équations, dont la solution générale dépend de 18 fonctions arbitraires de 3 arguments;

2° Des systèmes possibles de 16 équations dont la solution générale dépend de 16 fonctions arbitraires de 3 arguments;

3° Le système de M. Einstein et un autre de 22 équations (contenant deux constantes arbitraires), dont la solution générale dépend de 12 fonctions arbitraires de 3 arguments.

Le dernier système fournit en général une Physique *irréversible* (On se reportera à VII du 3 décembre, à VII N, à IX du 13 décembre et à [13]).

XXII

Le Chesnay (Seine et Oise)
27 avenue de Montespan
le 19 janvier 1930

Cher et illustre Maître,

J'ai été très occupé cette semaine et c'est pour cela que je ne vous ai pas écrit plus tôt au sujet de l'indice de généralité du système $R_{ik} = 0$.

Essayons d'abord de le déterminer par une méthode élémentaire qui malheureusement ne suffit pas. Le système considéré est un système d'équations aux dérivées partielles du second ordre par rapport aux fonctions $g_{\alpha\beta}$; on peut le ramener à un système du premier ordre par rapport aux fonctions

$$g_{\alpha\beta\gamma} = \frac{\partial g_{\alpha\beta}}{\partial x^{\gamma}}.$$

Les équations sont résolubles par rapport aux $\dfrac{\partial g_{\alpha\beta 1}}{\partial x^4}$, $\dfrac{\partial g_{\alpha\beta 2}}{\partial x^4}$, $\dfrac{\partial g_{\alpha\beta 3}}{\partial x^4}$ puisqu'elles contiennent

$$\frac{\partial g_{\alpha\beta 1}}{\partial x^4} - \frac{\partial g_{\alpha\beta 4}}{\partial x^1} = 0, \dots$$

Quant aux dérivées $\dfrac{\partial g_{\alpha\beta 4}}{\partial x^4}$, au nombre de 10, il n'y a que 6 équations pour les résoudre au moyen des dérivées prises par rapport à x^1, x^2, x^3. Cela correspond à ce fait que le choix des variables étant arbitraire, quatre des $g_{\alpha\beta}$ sont arbitraires. Pour fixer les idées on peut en effet toujours s'arranger pour que l'on ait

$$g_{44} = 1, \; g_{14} = 0, \; g_{24} = 0, \; g_{34} = 0$$

(la variable x^4 est alors solution de l'équation $g^{\rho\sigma} \dfrac{\partial \phi}{\partial x^{\rho}} \dfrac{\partial \phi}{\partial x^{\sigma}} = 1$, et les courbes $x^1 = c^{\text{te}}$, $x^2 = c^{\text{te}}$, $x^3 = c^{\text{te}}$ sont les trajectoires orthogonales des variétés $x^4 = c^{\text{te}}$).

A XXII

Le Chesnay (Seine et Oise)
27 avenue de Montespan
19 January 1930

Cher et illustre Maître,

I have been very busy this week and that's why I didn't write you sooner about the generality index of the system $R_{ik} = 0$.

Let us first try to determine it by an elementary method that unfortunately, will not be sufficient. The system under consideration is a system of second order partial differential equations with respect to the functions $g_{\alpha\beta}$; one can make it into a first order system with respect to the functions

$$g_{\alpha\beta\gamma} = \frac{\partial g_{\alpha\beta}}{\partial x^\gamma}.$$

The equations are solvable with respect to the $\frac{\partial g_{\alpha\beta 1}}{\partial x^4}, \frac{\partial g_{\alpha\beta 2}}{\partial x^4}, \frac{\partial g_{\alpha\beta 3}}{\partial x^4}$ since the former contain

$$\frac{\partial g_{\alpha\beta 1}}{\partial x^4} - \frac{\partial g_{\alpha\beta 4}}{\partial x^1} = 0, \ldots$$

As to the derivatives $\frac{\partial g_{\alpha\beta 4}}{\partial x^4}$, 10 in number, there are only 6 equations to resolve them by means of the derivatives with respect to x^1, x^2, x^3. This corresponds to the fact that, the choice of the coordinates being arbitrary, four of the $g_{\alpha\beta}$ are arbitrary. To fix ideas, one can always arrange things so as to have

$$g_{44} = 1, \ g_{14} = 0, \ g_{24} = 0, \ g_{34} = 0.$$

(The variable x^4 is then a solution for the equations $g^{\rho\sigma} \frac{\partial \phi}{\partial x^\rho} \frac{\partial \phi}{\partial x^\sigma} = 1$, and the curves $x^1 = $ const, $x^2 = $ const, $x^3 = $ const. are the trajectories orthogonal to the manifolds $x^4 = $ const.).

133

Jusqu'à présent tout marche bien. Voyons maintenant les équations du système qui ne contiennent pas les dérivées $\frac{\partial g_{\alpha\beta\gamma}}{\partial x^4}$. Il faut voir combien il y a de dérivées $\frac{\partial g_{\alpha\beta\gamma}}{\partial x^3}$ qui peuvent être obtenues au moyen des dérivées prises par rapport à x^1 et x^2. Il y a déjà les $\frac{\partial g_{\alpha\beta 1}}{\partial x^3} = \frac{\partial g_{\alpha\beta 3}}{\partial x^1}$ et les $\frac{\partial g_{\alpha\beta 2}}{\partial x^3} = \frac{\partial g_{\alpha\beta 3}}{\partial x^2}$. Reste les 20 dérivées $\frac{\partial g_{\alpha\beta 3}}{\partial x^3}$ et $\frac{\partial g_{\alpha\beta 4}}{\partial x^3}$. Nous pouvons déjà éliminer les 8 dérivées $\frac{\partial g_{\alpha 4 3}}{\partial x^3}$ et $\frac{\partial g_{\alpha 4 4}}{\partial x^3}$ puisque $g_{\alpha 4} = 0$ ou 1. Reste donc 12 dérivées. Or pour les obtenir on n'a que 4 équations; donc $12 - 4 = 8$ d'entre elles peuvent être prises arbitrairement; donc 8 des fonctions $g_{\alpha\beta\gamma}$ peuvent être prises arbitrairement dans la section $x^4 = c^{te}$. Il semblerait donc $\mathscr{I} = 8$.

Mais il faut faire attention que nous pouvons, grâce à l'arbitraire du choix des variables indépendantes, particulariser encore nos coefficients $g_{\alpha\beta}$. En particulier on peut, pour $x^4 = 0$, réduire la forme

$$g_{11}(dx^1)^2 + 2g_{12}dx^1dx^2 + ... + g_{33}(dx^3)^2$$

à

$$(dx^3)^2 + g_{11}(dx^1)^2 + ... + g_{22}(dx^2)^2,$$

c'est-à-dire *on peut supposer*, pour $x^4 = 0$,

$$g_{33} = 1, \; g_{13} = 0, \; g_{23} = 0;$$

cela supprime les 3 fonctions g_{133}, g_{233}, g_{333} qui sont nulles pour $x^4 = 0$. Cela réduit donc de 3 unités l'indice de généralité, qui serait par suite égal à 5 et non à 8.

Mais je ne sais pas si *en étant plus habile*, je n'arriverais pas à une réduction plus forte.

Pour voir les choses d'une manière indépendante de mon habileté, je remarque que ce qui caractérise complètement l'espace de Riemann, c'est la suite de ses invariants différentiels, dont les premiers sont les $R_i{}^j{}_{kh}$ puis leurs dérivées covariantes et ainsi de suite. Prenons alors une section $x^4 = 0$ ou une autre section quelconque à 3 dimensions. Attachons à chaque point de l'espace un 4-Bein rectangulaire tel que le 4e axe du 4-Bein soit normal à la section considérée. Les quantités R_{ijkh} ont leurs dérivées assujetties aux relations suivantes (identités de Bianchi)

$$R_{ij;khl} \equiv R_{ijkh;l} + R_{ijhl;k} + R_{ijlk;h} = 0. \tag{1}$$

So far everything is fine. Let us now look at the equations of the system that do not contain the derivatives $\frac{\partial g_{\alpha\beta\gamma}}{\partial x^4}$. One needs to know how many derivatives $\frac{\partial g_{\alpha\beta\gamma}}{\partial x^3}$ can be obtained by means of the x^1- and x^2-derivatives. We already have the $\frac{\partial g_{\alpha\beta 1}}{\partial x^3} = \frac{\partial g_{\alpha\beta 3}}{\partial x^1}$ and the $\frac{\partial g_{\alpha\beta 2}}{\partial x^3} = \frac{\partial g_{\alpha\beta 3}}{\partial x^2}$. There remain the 20 derivatives $\frac{\partial g_{\alpha\beta 3}}{\partial x^3}$ and $\frac{\partial g_{\alpha\beta 4}}{\partial x^3}$. We can eliminate the 8 derivatives $\frac{\partial g_{\alpha 43}}{\partial x^3}$ and $\frac{\partial g_{\alpha 44}}{\partial x^3}$ since $g_{\alpha 4} = 0$ or 1. So 12 derivatives are left over, and to obtain them we have only 4 equations, so $12 - 4 = 8$ of them can be taken arbitrarily; hence, 8 of the functions $g_{\alpha\beta\gamma}$ can be taken arbitrarily in the section $x^4 = $ const. It would seem that $\mathscr{I} = 8$.

But one should remember that, owing to the arbitrariness of the choice of the independent variables, we can specialize still further our coefficients $g_{\alpha\beta}$. In particular, for $x^4 = 0$, one can reduce the form

$$g_{11}(dx^1)^2 + 2g_{12}dx^1 dx^2 + \dots + g_{33}(dx^3)^2$$

to

$$(dx^3)^2 + g_{11}(dx^1)^2 + \dots + g_{22}(dx^2)^2;$$

that is, *one can assume that, for $x^4 = 0$,*

$$g_{33} = 1, \; g_{13} = 0, \; g_{23} = 0.$$

This eliminates the 3 functions $g_{133}, g_{233}, g_{333}$ that are zero for $x^4 = 0$. That reduces the generality index by 3 units that as a result, should be equal to 5 and not to 8.

But I don't known if, *by being more clever,* I couldn't achieve a greater reduction.

To see things independently of my cleverness, I note that what completely characterizes a Riemannian space is the sequence of its differential invariants, the first of which are the $R_i{}^j{}_{kh}$, then their covariant derivatives, and so on. Let us then take a section $x^4 = 0$ or any other 3-dimensional section. Let us attach an orthogonal 4-Bein to each point of space such that the 4th axis of the 4-Bein is normal to the section. The derivatives of the R_{ijkh} are subject to the following relations (Bianchi identities)

$$R_{ij;khl} \equiv R_{ijkh;l} + R_{ijhl;k} + R_{ijlk;h} = 0. \tag{1}$$

135

Si nous supposons que l'espace considéré soit un de ceux qui satisfont aux relations $R_{ik} = 0$, les relations (1) sont linéaires par rapport aux dérivées des 10 quantités R_{ijkh} restantes. Elles se réduisent dans ces conditions à 16: elles sont en effet au nombre de $6 \times 4 = 24$; mais on a identiquement

$$R_{i1,234} + R_{i2,314} + R_{i3,124} + R_{i4,132} \equiv 0, \quad (i = 1,2,3,4)$$
$$R_{23,231} + R_{24,241} + R_{34,341} \equiv 0,$$
$$R_{13,132} + R_{14,142} + R_{34,342} \equiv 0,$$
$$R_{12,123} + R_{14,143} + R_{24,243} \equiv 0,$$
$$R_{12,124} + R_{13,134} + R_{23,234} \equiv 0.$$

Des 16 équations indépendantes (1), 10 sont résolubles par rapport aux dérivées (prises par rapport à 4) des 10 R_{ijkh} indépendantes. Il en reste 6 ne contenant pas les dérivées par rapport à 4: ce sont manifestement les 6 relations

$$R_{ij,123} = 0;$$

elles sont résolubles par rapport aux dérivées, prises par rapport à 3, de 6 des quantités R_{ijkh}, à savoir $R_{12,12}$, $R_{12,13}$, $R_{12,14}$, $R_{12,23}$, $R_{12,24}$, $R_{12,34}$, et ces 6 quantités sont effectivement indépendantes. *Par conséquent 4 seulement des R_{ijkh} peuvent être prises arbitrairement dans la section $x^4 = 0$.* Si donc l'on ne porte son attention que sur les propriétés *indépendantes du choix des variables*, l'indice de généralité est effectivement 4, et non pas 5, ni 8.

Ce que je désigne par dérivée par rapport à 3, c'est la dérivée invariante dans la direction du 3e axe du 4-Bein.

Le même raisonnement montrerait que, dans le cas de n dimensions, l'indice de généralité des espaces satisfaisant à $R_{ik} = 0$ est $n(n-3)$ tandis que la méthode élémentaire indiquée tout d'abord donne seulement $\mathscr{I} \leqslant n^2 - 3n + 1$.

Si maintenant nous venons au système

$$R_{ik} - \frac{1}{2} g_{ik} R = -T_{ik},$$

$$\frac{\partial f_{ik}}{\partial x^k} = 0,$$

$$\frac{\partial f^{ik} \sqrt{-g}}{\partial x^k} = 0,$$

If we assume that the space satisfies the relations $R_{ik} = 0$, the relations (1) are linear with respect to the derivatives of the 10 remaining quantities R_{ijkh}. Under such conditions, they reduce to 16: in fact, there are $6 \times 4 = 24$ of them, but one has identically,

$$R_{i1,234} + R_{i2,314} + R_{i3,124} + R_{i4,132} \equiv 0, \ (i = 1,2,3,4)$$
$$R_{23,231} + R_{24,241} + R_{34,341} \equiv 0,$$
$$R_{13,132} + R_{14,142} + R_{34,342} \equiv 0,$$
$$R_{12,123} + R_{14,134} + R_{24,243} \equiv 0,$$
$$R_{12,124} + R_{13,134} + R_{23,234} \equiv 0.$$

Among the 16 independent equations (1), 10 can be solved with respect to the 4-derivatives of the 10 independent R_{ijkh}. There remain 6 without the 4-derivatives: clearly they are the 6 relations

$$R_{ij,123} = 0,$$

which can be solved with respect to the 3-derivatives of 6 of the quantities R_{ijkh}, namely $R_{12,12}$, $R_{12,13}$, $R_{12,14}$, $R_{12,23}$, $R_{12,24}$, $R_{12,34}$ and these 6 quantities are actually independent. *Consequently, only 4 of the R_{ijkh} can be taken arbitrarily in the cross-section $x^4 = 0$.* So, if we fix our attention only on the properties that are *independent of the choice of variables*, the generality index is actually 4 and not 5, or 8.

By 3-derivative, I mean the invariant derivative in the direction of the third axis of the 4-Bein.

The same reasoning would show that, in the case of n dimensions, the generality index of the spaces satisfying $R_{ik} = 0$ is $n(n - 3)$, while the elementary method outlined above gives only $\mathscr{I} \leqslant n^2 - 3n + 1$.

If we now consider to the system

$$R_{ik} - \frac{1}{2} g_{ik} R = -T_{ik},$$

$$\frac{\partial f_{ik}}{\partial x^k} = 0,$$

$$\frac{\partial f^{ik}\sqrt{-g}}{\partial x^k} = 0,$$

137

nous trouvons $\mathscr{I} = 4 + 4 = 8$. Seulement il faut remarquer que ce système ne donne le champ que dans le vide, puisque vous excluez a priori les singularités. En tous cas les raisons que vous m'indiquez de préférer le nouveau système à l'ancien me paraissent tout à fait naturelles, et elles sont aussi valables en ce qui concerne le système de 22 équations avec les $R_{ik} = 0$, système dont l'indice de généralité est bien 12.

Francis Perrin, le fils de Jean Perrin, m'a communiqué l'autre jour une remarque intéressante. En modifiant votre système de la manière suivante:

$$\Lambda^{\mu}_{\alpha\beta;\mu} = 0,$$
$$G^{\beta}_{\underline{\alpha}} = kg^{\alpha\beta},$$

k étant une *constante* arbitraire, on obtient encore un système en involution avec le même degré de généralité. Ce nouveau système admet une solution sans singularité dans l'espace sphérique (avec temps indéfini). On peut du reste supposer $k = 1$ sans restreindre la généralité (cela revient à changer l'unité de longueur). Cela m'a conduit à la remarque suivante: l'hypothèse que les équations cherchées contiennent les $\Lambda^{\gamma}_{\alpha\beta}$ au second degré d'une manière homogène revient à l'hypothèse que *les équations cherchées sont invariantes par un changement de l'unité de longueur*. (Pensez-vous qu'il y ait a priori une raison de penser qu'il existe dans la nature une unité de longueur privilégiée) (ou plutôt une unité d'*intervalle* privilégiée)? L'idée de F. Perrin serait peut-être alors à retenir [1].

Veuillez agréer, cher et illustre Maître, l'expression de mes sentiments bien dévoués.

E. Cartan

---

1. Le Professeur Francis Perrin consulté à ce sujet nous a écrit « Je me souviens avoir moi-même travaillé dans ce domaine à la suite des exposés que nous avait faits E. Cartan auquel j'avais communiqué certaines remarques que m'avaient inspirées ces exposés, mais je n'ai fait aucune publication à ce sujet » (Lettre du 18.10.78). Le professeur Perrin nous a également communiqué quelques pages de calculs qu'il avait écrites à l'époque sur le cas tridimensionnel. Il s'agit d'un parallélisme absolu de la sphère à trois dimensions qui satisfait aux équations d'Einstein modifiées
$$F^{\alpha\beta} = 0; \ G^{\alpha\beta} = kg^{\alpha\beta}.$$

we find $\mathcal{I} = 4 + 4 = 8$. But one should note that this system gives the field only in vacuum, since you *a priori* exclude singularities. In any case, the reasons you give for preferring the new system to the old one seem quite natural to me, and they are also valid for the system of 22 equations with $R_{ik} = 0$, for which the generality index is 12.

Francis Perrin, the son of Jean Perrin, made an interesting remark to me the other day. By modifying your system in the following way

$$\Lambda^{\mu}_{\alpha\beta;\mu} = 0,$$

$$G^{\beta}_{\underline{\alpha}} = kg^{\alpha\beta},$$

k being an arbitrary *constant*, one still has a system in involution with the same degree of generality. This new system admits a singularity-free solution in spherical space (with indefinite time). Moreover, one can assume $k = 1$ without loss of generality (this amounts to changing the unit of length). This leads me to the following remark: the assumption that the equations one is looking for are quadratic and homogeneous in the $\Lambda^{\gamma}_{\alpha\beta}$ is equivalent to the assumption that *these equations are invariant for a change of the unit of length*. (Do you think that there is, *a priori*, any reason to think that there exists, in nature, a privileged unit of length, or rather a privileged unit of *interval*)? Perhaps then, Perrin's idea would be worth remembering [1]...

E. Cartan

De plus nous avons reçu également du Professeur Perrin communication d'une lettre d'Einstein, datée du 10 février 1930, qui fait sans doute suite à la lettre de Cartan du 7 février et dans laquelle Einstein souligne l'intérêt mathématique du résultat et la nécessité qu'il y a de s'assurer de l'homogénéité de l'espace des repères et de l'absence de singularités. Nous n'avons pas d'autres indications quant à la solution à laquelle Cartan fait allusion. Voir aussi lettre XXX.

XXIII

21-I-30

Verehrter Herr Cartan!

Ich empfinde es selbst als eine Vermessenheit, Ihnen auf mathematischem Gebiete zu opponieren. Aber es hilft nichts. Ich kann Ihre beiden Argumente zugunsten von $\mathscr{I} = 4$ bei $R_{ik} = 0$ *nicht* begreifen.

Zu Ihrem ersten Beweise. Sie fragen nach $\mathscr{I}$ für dasjenige Gleichungssystem, welches aus dem invarianten durch die Spezialisierung

$$g_{14} = g_{24} = g_{34} = 0, \quad g_{44} = 1,$$

entsteht. Dies begreife ich, da es einer möglichen, allgemein durchführbaren Spezialieserung der Koordinatenwahl entspricht.

Dann aber müssen wir einfach fragen: Wieviele Funktionen dürfen wir dem Schnitte $x^4 =$ konst. noch frei vorschreiben. Wir dürfen aber nicht fragen: wieviele von diesen Funktionen können wir *bei bestimmter Koordinatenwahl auf dieser Fläche* noch frei wählen. Denn die Antwort auf diese Frage liefert nicht die uns interessierende Grösse $\mathscr{I}$.

Es scheint mir logisch, so zu sagen:
Das Gleichungssystem lautet

$$R_{ik} = 0,$$

$$g_{14} = 0, \ g_{24} = 0, \ g_{34} = 0, \ g_{44} = 1.$$

Welches ist das zugehörige $\mathscr{I}$? Analog ist die Frage auch im Falle meines Gleichungssystems gestellt.

Der andere Beweis ist mir aus folgendem Grunde nicht verständlich. Wir müssen uns darüber entscheiden, ob wir die

$$R^i_{jkl} \text{ oder } R_{ijkl}$$

als die Feldvariabeln ansehen. Die Zahl der ersten ist

$$6 \cdot 16 - 1$$

die der zweiten $\dfrac{6 \cdot 7}{2} - 1$

wegen der zusätzlichen Symmetrien.

140

A XXIII

21.1.30

Dear M. Cartan,

I feel it presumptions in myself to oppose you on a mathematical issue; but it can't be helped. I *cannot* understand your two arguments in favour of $\mathscr{I} = 4$ for $R_{ik} = 0$.

In your first proof you ask for $\mathscr{I}$ in the case of that system of equations arising from invariants through the specialization

$$g_{14} = g_{24} = g_{34} = 0 \quad g_{44} = 1.$$

I understand this, since it corresponds to a possible, generally feasible, special choice of coordinates.

Then we ought simply to ask: how many functions may we still freely prescribe on the cross-section $x^4 = $ const.? But we may not ask: how many of these functions can we still freely choose *for a fixed choice of coordinates on this surface?* For the answer to this question does not provide the value of $\mathscr{I}$ which interests us.

To me it seems logical to say: the system of equations reads

$$R_{ik} = 0,$$

$$g_{14} = 0, \; g_{24} = 0, \; g_{34} = 0, \; g_{44} = 1.$$

What is the corresponding $\mathscr{I}$? The analogous question applies in the case of my system of equations.

I don't understand the other proof for the following reason. We must decide whether we view

$$R^{i}{}_{jkl} \text{ or } R_{ijkl}$$

as the field variables. In the first case their number is

$$6 \cdot 16 - 1$$

and the second case only

$$\frac{6 \cdot 7}{2} - 1$$

Aus den R^i_{jkl} kann ich nun die R_{ik} durch Kontraktion bilden, nicht aber aus R_{iklm}, weil ja dann die g_{ik} intervenieren, was ja gerade vermieden werden sollte.

Also haben Sie *noch mehr* Geduld und Mitleid mit einem armen Physiker, den die Not in diese schwierige Gefilde geführt hat.

Herzlich grüsst Sie

Ihr

A. Einstein

142

because of additional symmetries. Now, I can form the R_{ik} by contraction from the R^i_{jkl} but not from the R_{iklm}, because then the g_{ik} enter in, which is precisely what was to be avoided.

Thus you must have *still more* patience and compassion for a poor physicist whose destiny has led him to these troubled pastures.

Kind regards.

Yours,

A. Einstein

XXIV

Le Chesnay (S. et O.)
27 avenue de Montespan,
le 26 janvier 1930

Cher et illustre Maître,

J'ai bien reçu votre lettre du 21. Ma seconde démonstration est en effet très critiquable et elle envisage les choses d'une manière un peu trop sommaire. J'avais espéré pouvoir me passer d'une démonstration tout à fait générale, mais qui demande des explications assez délicates; je me vois obligé de sortir cette machine de guerre, qui n'est pas très commode à manier! Avant de parler de cette démonstration générale, je viens à votre critique de ma première démonstration.

Sur ce point, je ne suis pas encore d'accord avec vous. Vous admettez la particularisation

$$g_{14} = g_{24} = g_{34} = 0, \ g_{44} = 1$$

et il s'agit alors de voir ce qui se passe dans la section $x^4 = C^{te}$. Mais, à mon avis, il s'agit de voir ce qui s'y passe, *indépendamment du choix des variables dans cette section.*

Prenons les choses de plus haut. Nous cherchons, si je ne me trompe (je pense que c'est aussi la question que vous voulez résoudre), le degré d'indétermination des espaces riemanniens satisfaisant aux conditions $R_{ik} = 0$. *Nous regardons donc comme identiques deux de ces espaces si par un changement convenable de variables, les $g_{\alpha\beta}$ du premier peuvent être identifiés aux $g_{\alpha\beta}$ du second.* Or prenons un espace riemannien quelconque; nous pouvons toujours faire un changement de variables tel qu'on ait identiquement

$$g_{14} = 0, \ g_{24} = 0, \ g_{34} = 0, \ g_{44} = 1:$$

sur ce point nous sommes d'accord. Mais dans cet espace nous pouvons, parmi tous les changements de variables satisfaisant à la condition précédente, en choisir une infinité de telle sorte que, *pour $x^4 = 0$,* on ait

$$g_{33} = 1, \ g_{13} = g_{23} = 0;$$

144

A XXIV

Le Chesnay (S. et O.)
27 avenue de Montespan,
26 January 1930

Cher et illustre Maître,

I have received your letter of the 21st. My second proof is very subject to criticism indeed and it considers things in too sketchy a way. I had hoped to manage to avoid a completely general proof that demands very delicate explanations, but I am compelled to bring out this war machine that is not very easy to handle! Before talking about this general proof, I shall touch on your criticism of my first proof.

On this point, I don't yet agree with you. You allow the specialization

$$g_{14} = g_{24} = g_{34} = 0, \; g_{44} = 1,$$

and the question is then to see what happens in the section $x^4 = $ const. But, in my opinion, the question is what happens *independently of the choice of variables in this section.*

Let us look at things from a wider vantage point. If I'm not mistaken (and I think it's also the question you want to solve), we are looking for the degree of indetermination of the Riemannian spaces satisfying the conditions $R_{ik} = 0$. *Hence we consider two of these spaces to be identical if by a suitable change of variables, the $g_{\alpha\beta}$ of the first one can be identified with the $g_{\alpha\beta}$ of the second one.* Now let us take any Riemannian space; we can always perform a change of variables such that one has, identically,

$$g_{14} = 0, \; g_{24} = 0, \; g_{34} = 0, \; g_{44} = 1.$$

On this we agree. But in this space, among all the changes of variables satisfying the previous condition, we can choose an infinity of them in such a way that *for $x^4 = 0$,* one has

$$g_{33} = 1, \; g_{13} = g_{23} = 0.$$

145

c'est une chose qui n'est pas non plus contestable. Je suis donc sûr d'obtenir *tous* les espaces riemanniens à $R_{ik} = 0$ en formant le système d'équations

$$R_{ik} = 0, \quad g_{14} = 0, \quad g_{24} = 0, \quad g_{34} = 0, \quad g_{44} = 1 \tag{1}$$

et en cherchant *dans la section particulière* $x^4 = 0$, les solutions à 3 dimensions qui satisfont aux conditions supplémentaires

$$g_{13} = 0, \quad g_{23} = 0, \quad g_{33} = 1. \tag{2}$$

Cette solution à 3 dimensions détermine une solution complète du système (1) et cette solution me fournit certainement *tous* les espaces riemanniens satisfaisant à $R_{ik} = 0$; ou du moins je puis affirmer que tout espace riemannien satisfaisant à $R_{ik} = 0$ est *applicable* sur l'un des espaces riemanniens que j'ai trouvés (et en réalité sur une infinité). Le degré de généralité cherché ne dépasse donc pas le degré de généralité de la solution à 3 dimensions ($x^4 = 0$) du système (1) et (2). La seule chose que je ne puisse pas affirmer c'est qu'il lui est égal; il lui est *peut-être* inférieur: $\mathscr{I} \leqslant 5$.

En résumé il me semble qu'on peut seulement se poser deux problèmes.

1. Quel est l'indice de généralité du système

$$R_{ik} = 0$$

considéré comme admettant les $g_{\alpha\beta}$ pour fonctions inconnues? La réponse est: la solution générale dépend de 4 fonctions arbitraires de 4 variables.

2. Quel est l'indice de généralité du même système, en ne regardant pas comme distinctes deux solutions qui peuvent se déduire l'une de l'autre par un changement de variables? La réponse est: la solution générale dépend de $\mathscr{I} \leqslant 5$ fonctions arbitraires de 3 variables.

Je ne vois pas très bien quel pourrait être le problème intermédiaire auquel vous devez penser, si vos critiques sont fondées.

J'arrive maintenant à la méthode générale permettant de trouver avec certitude l'indice d'indétermination, quand on regarde comme identiques deux solutions se réduisant l'une à l'autre par un changement de variables. Pour aujourd'hui je me bornerai au cas simple des espaces riemanniens *à parallélisme absolu*.

This is also an unquestionable fact. So, I'm sure to obtain *all* the Riemannian spaces with $R_{ik} = 0$ if I consider the system of equations

$$R_{ik} = 0, \ g_{14} = 0, \ g_{24} = 0, \ g_{34} = 0, \ g_{44} = 1, \tag{1}$$

and if I look, *in the particular section* $x^4 = 0$, for the 3-dimensional solutions satisfying the additional conditions

$$g_{13} = 0, \ g_{23} = 0, \ g_{33} = 1. \tag{2}$$

This 3-dimensional solution determines a complete solution of system (1) and this solution certainly gives me *all* the Riemannian spaces satisfying $R_{ik} = 0$; or at least I can assert that any Riemannian space satisfying $R_{ik} = 0$, is *applicable* on one of the Riemannian spaces I have found (and in fact on an infinity of them). The degree of generality one is looking for is, therefore, not larger than the degree of generality of the 3-dimensional solution ($x^4 = 0$) of the system (1) and (2). The only thing I cannot assert is that it is equal to the latter, *perhaps* it is smaller, $\mathscr{I} \leqslant 5$.

To summarize, it seems to me that only two problems can be posed.

1. What is the generality index of the system

$$R_{ij} = 0,$$

considered as having the $g_{\alpha\beta}$ as unknown functions? The answer is: The general solution depends on 4 arbitrary functions of 4 variables.

2. What is the generality index of the same system, when we are not considering as distinct two solutions that can be deduced from each other by a change of variables? The answer is: The general solution depends on $\mathscr{I} \leqslant 5$ arbitrary functions of 3 variables.

I don't see very clearly what the intermediate problem could be that you must be thinking of if your criticisms are to be valid.

I now consider the general method allowing us to find with certainty, the indeterminancy index when one considers two solutions to be identical when they can be reduced to each other by a change of variables. I shall restrict myself for now to the simple case of Riemannian spaces with *absolute parallelism*.

Prenons un espace riemannien à parallélisme absolu, satisfaisant par exemple à votre système. Attachons aux différents points de cet espace des 4-Bein parallèles entre eux (un 4-Bein à chaque point), et désignons par Λ_{ij}^k les composantes de la torsion rapportées à ce 4-Bein, par $\Lambda_{ij;h}^k$ la dérivée (ordinaire) de Λ_{ij}^k prise dans la direction du h^e Bein.

Je pars de la remarque suivante. Prenons quatre des Λ_{ij}^k qui soient des fonctions indépendantes de x^1, x^2, x^3, x^4. Désignons-les pour abréger par $\Lambda^{(1)}$, $\Lambda^{(2)}$, $\Lambda^{(3)}$, $\Lambda^{(4)}$. Les quantités Λ_{ij}^k et $\Lambda_{ij;h}^k$ sont, dans mon espace riemannien considéré, des fonctions déterminées de $\Lambda^{(1)}$, $\Lambda^{(2)}$, $\Lambda^{(3)}$, $\Lambda^{(4)}$:

$$\Lambda_{ij}^k = \phi_{ij}^k(\Lambda^{(1)}, \Lambda^{(2)}, \Lambda^{(3)}, \Lambda^{(4)}),$$
$$\Lambda_{ij;h}^k = \phi_{ijh}^k(\Lambda^{(1)}, \Lambda^{(2)}, \Lambda^{(3)}, \Lambda^{(4)}).$$

Je dis que *si pour deux espaces différents* (à parallélisme absolu) *les fonctions ϕ_{ij}^k et ϕ_{ijh}^k sont les mêmes*, ces deux espaces sont identiques (c'est-à-dire ne diffèrent que par le choix des variables). En effet, soient x^α et $h_{s\alpha}(x)$ les variables et les fonctions $h_{s\alpha}$ de l'espace, soient $\bar{x}^\alpha$ et $\bar{h}_{s\alpha}$ les quantités correspondantes du second. Établissons entre les deux espaces la correspondance ponctuelle définie par les 4 relations

$$\Lambda^{(1)} = \bar{\Lambda}^{(1)}, \ \Lambda^{(2)} = \bar{\Lambda}^{(2)}, \ \Lambda^{(3)} = \bar{\Lambda}^{(3)}, \ \Lambda^{(4)} = \bar{\Lambda}^{(4)}; \tag{3}$$

on aura alors, par cette correspondance ponctuelle,

$$\Lambda_{ij}^k = \bar{\Lambda}_{ij}^k; \ \Lambda_{ij;h}^k = \bar{\Lambda}_{ij;h}^k \ ;$$

en particulier, on aura

$$\Lambda^{(i)}_{;k} = \bar{\Lambda}^{(i)}_{;k} \ (i,k = 1,2,3,4). \tag{4}$$

Or on a

$$d\Lambda^{(i)} = \Lambda^{(i)}_{;s} h_{s\alpha} dx^\alpha,$$

et

$$d\bar{\Lambda}^{(i)} = \bar{\Lambda}^{(i)}_{;s} \bar{h}_{s\alpha} d\bar{x}^\alpha.$$

Par suite la correspondance ponctuelle que nous avons établie entre les deux espaces par les relations (3) nous donne, en tenant compte de (4),

$$h_{s\alpha} dx^\alpha \equiv \bar{h}_{s\alpha} d\bar{x}^\alpha \ ;$$

les deux espaces sont donc identiques (au sens que nous donnons à ce mot).

Let us take a Riemannian space with absolute parallelism which satisfies for instance your system. Let us attach to the points of that space 4-Beins which are parallel to each other (one 4-Bein at each point), and let us call Λ_{ij}^k the components of the torsion in the 4-Bein and $\Lambda_{ij;h}^k$ the (ordinary) derivative of Λ_{ij}^k in the direction of the hth Bein.

I start off with the following remark. Take four of the Λ_{ij}^k that are independent functions of x^1, x^2, x^3, x^4 and call them for short, $\Lambda^{(1)}$, $\Lambda^{(2)}$, $\Lambda^{(3)}$, $\Lambda^{(4)}$. In my Riemannian space, the quantities Λ_{ij}^k and $\Lambda_{ij;h}^k$ are well defined functions of $\Lambda^{(1)}$, $\Lambda^{(2)}$, $\Lambda^{(3)}$, $\Lambda^{(4)}$.

$$\Lambda_{ij}^k = \phi_{ij}^k(\Lambda^{(1)}, \Lambda^{(2)}, \Lambda^{(3)}, \Lambda^{(4)}),$$
$$\bar{\Lambda}_{ij;h}^k = \phi_{ijh}^k(\Lambda^{(1)}, \Lambda^{(2)}, \Lambda^{(3)}, \Lambda^{(4)}).$$

I say that *if for two different spaces* (with absolute parallelism) *the functions ϕ_{ij}^k and ϕ_{ijh}^k are the same*, the two spaces are identical (that is, differ only by the choice of variables). Indeed, let x^α and $h_{s\alpha}(x)$ be the variables and the functions $h_{s\alpha}$ of the space, and $\bar{x}^\alpha$ and $\bar{h}_{s\alpha}$ be the corresponding quantities of the second one. Let us establish between the two spaces the point-to-point correspondence defined by the 4 relations

$$\Lambda^{(1)} = \bar{\Lambda}^{(1)}, \ \Lambda^{(2)} = \bar{\Lambda}^{(2)}, \ \Lambda^{(3)} = \bar{\Lambda}^{(3)}, \ \Lambda^{(4)} = \bar{\Lambda}^{(4)}. \qquad (3)$$

Then, by this point-to-point correspondence, we have

$$\Lambda_{ij}^k = \bar{\Lambda}_{ij}^k; \ \Lambda_{ij;h}^k = \bar{\Lambda}_{ij;h}^k \ ;$$

in particular, we have

$$\Lambda^{(i)}{}_{;k} = \bar{\Lambda}^{(i)}{}_{;k} \ (i,k = 1,2,3,4). \qquad (4)$$

But we have

$$d\Lambda^{(i)} = \Lambda^{(1)}{}_{;s}h_{s\alpha}dx^\alpha,$$

and

$$d\bar{\Lambda}^{(1)} = \bar{\Lambda}^{(i)}{}_{;s}\bar{h}_{s\alpha}d\bar{x}^\alpha.$$

As a result, the correspondence that we have established between the two spaces by the relations (3) gives us, because of (4)

$$h_{s\alpha}dx^\alpha \equiv \bar{h}_{s\alpha}d\bar{x}^\alpha,$$

and so the two spaces are identical (in the meaning we give to this word).

149

Ce qui caractérise essentiellement un espace riemannien à parallélisme absolu, ce sont donc les fonctions ϕ_{ij}^k *et* ϕ_{ijh}^k *des quatre variables* $\Lambda^{(1)}$, $\Lambda^{(2)}$, $\Lambda^{(3)}$, $\Lambda^{(4)}$. La question qui se pose est alors la suivante. Comment former le système différentiel qui définit ces fonctions et comment trouver l'indice de généralité de ce système? Voici comment on peut y parvenir.

Partons pour cela des équations du champ (avec fonctions inconnues $h_{s\alpha}$ et $\Lambda_{\alpha\beta}^\gamma$), mais n'écrivons pas celles qui définissent les $\Lambda_{\alpha\beta}^\gamma$ au moyen des dérivées partielles des $h_{s\alpha}$. Quant aux autres écrivons-les sous la forme qui fait intervenir les composantes Λ_{ij}^k rapportées à un système de 4-Bein parallèles, ainsi que leurs dérivées $\Lambda_{ij;h}^k$ prises dans la direction des axes du 4-Bein. Nous aurons les 38 équations

$$
\begin{cases}
H_{ijk}^l \equiv \Lambda_{ij;k}^l + \Lambda_{jk;i}^l + \Lambda_{ki;j}^l + \Lambda_{ij}^m\Lambda_{km}^l + \Lambda_{jk}^m\Lambda_{im}^l + \Lambda_{ki}^m\Lambda_{jm}^l = 0, \\
F_{ij} \equiv \Lambda_{ij;k}^k = 0, \\
G_i^j \equiv \Lambda_{ik;k}^j + \Lambda_{ik}^m\Lambda_{km}^j = 0.
\end{cases}
\tag{I}
$$

Rappelons que de ces 38 équations, 2 dont nous désignerons les premiers membres par A_1, A_2, ne contiennent que les $\Lambda_{ij;1}^k$ et $\Lambda_{ij;2}^k$; 12 autres indépendantes ne contiennent que les $\Lambda_{ij;1}^k$, $\Lambda_{ij;2}^k$ et $\Lambda_{ij;3}^k$; enfin les 24 autres sont linéairement inépendantes par rapport aux $\Lambda_{ij;4}^k$. Nous désignerons alors le système (I) par

$$
\begin{cases}
A_i = 0, \ i = 1, 2 \\
B_j = 0, \ j = 1, 2, ..., 12 \\
C_h = 0, \ h = 1, 2, ..., 24.
\end{cases}
\tag{I'}
$$

Nous allons maintenant prendre les $\Lambda_{ij;h}^k$ que nous désignerons par Λ_{ijh}^k, comme nouvelles fonctions inconnues. Elles satisfont aux équations

$$
\begin{cases}
L_{ijkh}^l \equiv \Lambda_{ijk;h}^l - \Lambda_{ijh;k}^l + \Lambda_{kh}^m\Lambda_{ijm}^l = 0, \\
H_{ijkh}^l \equiv \Lambda_{ijk;h}^l + \Lambda_{jki;h}^l + \Lambda_{kij;h}^l + \Lambda_{ij}^m\Lambda_{kmh}^l + ... = 0, \\
F_{ijh} \equiv \Lambda_{ijk;h}^k = 0, \\
G_{ih}^j \equiv \Lambda_{ikk;h}^j + \Lambda_{ikh}^m\Lambda_{km}^j + \Lambda_{ik}^m\Lambda_{kmh}^j = 0.
\end{cases}
\tag{II}
$$

Les équations (II) ne sont du reste pas toutes linéairement indépendantes, car elles sont manifestement liées par 16 relations identiques (en tenant compte de (I) et (II)).

The features that essentially characterize a Riemannian space with absolute parallelism i.e the function ϕ_{ij}^k and ϕ_{ijh}^k of the four variables $\Lambda^{(1)}$, $\Lambda^{(2)}$, $\Lambda^{(3)}$, $\Lambda^{(4)}$. The question is then the following: How to form the differential system defining these functions and how to find the generality index of this system? Here is how we do it.

Let us start from the field equations (with $h_{s\alpha}$ and $\Lambda_{\alpha\beta}^\gamma$ as unknown functions), but let us not write the equations defining the $\Lambda_{\alpha\beta}^\gamma$ by means of the partial derivatives of the $h_{s\beta}$. As for the others, let us write them in a form that involves the components Λ_{ij}^k in a system of parallel 4-Beins, equally their derivatives $\Lambda_{ij;h}^k$ in the direction of the axes of the 4-Bein. We then have the 38 equations,

$$
\begin{cases}
H_{ijk}^l \equiv \Lambda_{ij;k}^l + \Lambda_{jk;i}^l + \Lambda_{ki;j}^l + \Lambda_{ij}^m \Lambda_{km}^l + \Lambda_{jk}^m \Lambda_{im}^l + \Lambda_{ki}^m \Lambda_{jm}^l = 0, \\
F_{ij} \equiv \Lambda_{ij;k}^k = 0, \\
G_i^j \equiv \Lambda_{ik;k}^j + \Lambda_{ik}^m \Lambda_{km}^j = 0.
\end{cases}
\tag{I}
$$

Let us recall that in these 38 equations, two — the left hand sides of which will be denoted by A_1, A_2 — contain only the $\Lambda_{ij;1}^k$ and $\Lambda_{ij;2}^k$; 12 other independent ones contain only the $\Lambda_{ij;1}^k$, $\Lambda_{ij;2}^k$ and $\Lambda_{ij;3}^k$; finally the last 24 are linearly independent with respect to the $\Lambda_{ij;4}^k$. Then, we shall denote the system (I) by

$$
\begin{cases}
A_i = 0, \ i = 1, 2, \\
B_j = 0, \ j = 1, 2, ..., 12, \\
C_h = 0, \ h = 1, 2, ..., 24.
\end{cases}
\tag{I'}
$$

We are now going to take the $\Lambda_{ij;h}^k$ denoted by Λ_{ijh}^k, as new unknown functions. They satisfy the equations

$$
\begin{cases}
L_{ijkh}^l \equiv \Lambda_{ijk;h}^l - \Lambda_{ijh;k}^l + \Lambda_{kh}^m \Lambda_{ijm}^l = 0, \\
H_{ijkh}^l \equiv \Lambda_{ijk;h}^l + \Lambda_{jki;h}^l + \Lambda_{kij;h}^l + \Lambda_{ij}^m \Lambda_{kmh}^l + ... = 0, \\
F_{ijh} \equiv \Lambda_{ijk;h}^k = 0, \\
G_{ih}^j \equiv \Lambda_{ikk;h}^j + \Lambda_{ikh}^m \Lambda_{km}^j + \Lambda_{ik}^m \Lambda_{kmh}^j = 0.
\end{cases}
\tag{II}
$$

However, equations (II) are not all linearly independent, for they are obviously related by 16 identities (on account of (I) and (II)).

Nous pouvons ordonner le système (II) comme nous avons ordonné le système (I). Nous aurons les suites d'équations indépendantes

$$(\text{II}')\begin{cases} 1° \ A_{i1} = 0, \ B_{j1} = 0, \ C_{k1} = 0 & (38) \\[4pt] 2° \ A_{i2} = 0, \ B_{j2} = 0, \ C_{k2} = 0, \ L^l_{ij12} = 0 & (62) \\[4pt] 3° \ B_{j3} = 0, \ C_{k3} = 0, \ L^l_{ij13} = 0, \ L^l_{ij23} = 0 & (84) \\[4pt] 4° \ C_{k4} = 0, \ L_{ij14} = 0, \ L_{ij24} = 0, \ L_{ij34} = 0 & (96) \end{cases}$$

Les équations 1° ne contiennent que des dérivées $\Lambda^l_{ijk;1}$; les équations 2° ne contiennent que des dérivées $\Lambda^l_{ijk;1}$ et $\Lambda^l_{ijk;2}$; les équations 3° ne contiennent que des dérivées $\Lambda^l_{ijk;1}, ..., \Lambda^l_{ijk;3}$. Les équations 4° sont manifestement indépendantes par rapport aux 24 × 4 dérivées $\Lambda^k_{ijh;4}$. Quant aux équations 3°, elles ne permettent de déterminer que 24 × 4 − 12 dérivées $\Lambda^k_{ijh;3}$ au moyen des dérivées ;1 et ;2 et des 12 autres dérivées $\Lambda^k_{ijh;3}$. *Ce nombre 12 est le même que celui des dérivées $\Lambda^k_{ij;3}$ que les équations* $B_j = 0$ *ne permettaient pas de déterminer au moyen des $\Lambda^k_{ij;1}$ et des $\Lambda^k_{ij;2}$.*

Revenons maintenant à notre problème. Remplaçons dans les équations (I) et (II) les quantités $\Lambda^l_{ij;k}$ par $\dfrac{\partial \phi^l_{ij}}{\partial \Lambda^{(s)}}\Lambda^{(s)}{}_{;k}$ et les quantités $\Lambda^l_{ijk;h}$ par $\dfrac{\partial \phi^l_{ijk}}{\partial \Lambda^{(s)}}\Lambda^{(s)}{}_{;h}$. Il est à remarquer que les quantités $\Lambda^{(s)}{}_{;h}$ sont elles-mêmes 16 des fonctions ϕ^l_{ijk}. Par cette substitution les équations (I) et (II) deviennent un système d'équations aux dérivées partielles *ordinaires* aux fonctions inconnues ϕ^l_{ij} et ϕ^l_{ijk} de $\Lambda^{(1)}, \Lambda^{(2)}, \Lambda^{(3)}, \Lambda^{(4)}$. On a du reste

$$\phi^l_{ijk} = \frac{\partial \phi^l_{ij}}{\partial \Lambda^{(s)}} \Lambda^{(s)}{}_{;k}.$$

En définitive il y a 24 − 4 = 20 fonctions ϕ^l_{ij} et 96 fonctions ϕ^l_{ijk}.

On pourrait montrer facilement, mais ce n'est pas indispensable, que ce système d'équations aux dérivées partielles ordinaires est en involution. Les considérations développées sur les nombres respectifs des équations 1°, 2°, 3°, 4° du système (II′) nous montrent que si l'on ordonne les *nouvelles* équations (II) linéaires par rapport aux $\dfrac{\partial \phi_{ijk}}{\partial \Lambda^{(s)}}$, par rapport aux dérivées $\dfrac{\partial}{\partial \Lambda^{(1)}}, \dfrac{\partial}{\partial \Lambda^{(2)}}, \dfrac{\partial}{\partial \Lambda^{(3)}}, \dfrac{\partial}{\partial \Lambda^{(4)}}$, on arrivera aux mêmes nombres (dont la valeur est toujours la même quel que soit le choix non caractéristique des variables). Par suite

We can order the system (II) as we have ordered the system (1). We then have the following sequences of independent equations

$$
\left\{
\begin{array}{ll}
\text{1) } A_{i1} = 0, \; B_{j1} = 0, \; C_{k1} = 0 & (38) \\[4pt]
\text{2) } A_{i2} = 0, \; B_{j2} = 0, \; C_{k2} = 0, \; L^l_{ij12} = 0 & (62) \\[4pt]
\text{3) } B_{j3} = 0, \; C_{k3} = 0, \; L^l_{ij13} = 0, \; L^l_{ij23} = 0 & (84) \\[4pt]
\text{4) } C_{k4} = 0, \; L_{ij14} = 0, \; L_{ij24} = 0, \; L_{ij34} = 0 & (96)
\end{array}
\right. \quad \text{(II')}
$$

Equations 1) contain only the derivatives $\Lambda^l_{ijk,1}$; equations 2) only the derivatives $\Lambda^l_{ijk;1}$ and $\Lambda^l_{ijk;2}$; equations 3) only the derivatives $\Lambda^l_{ij;k1}$ ... $\Lambda^l_{ijk;3}$. Equations 4) are obviously independent with respect to the 24×4 derivatives $\Lambda^k_{ij;h4}$. As for equations 3), they allow us to determine only the $24 \times 4 - 12$ derivatives $\Lambda^k_{ijh;3}$ by means of the $;1$ and $;2$ derivatives and the other 12 derivatives $\Lambda^k_{ijh;3}$. *This number 12 is the same as the number of derivatives $\Lambda^k_{ij;3}$ that the equations $B_j = 0$ did not allow us to determine by means of the $\Lambda^k_{ij;1}$ and $\Lambda^k_{ij;2}$.*

Let us now return to our problem. We replace in equations (I) and (II), the quantities $\Lambda^l_{ij;k}$ by $\dfrac{\partial \phi^l_{ij}}{\partial \Lambda^{(s)}} \Lambda^{(s)}_{;k}$ and the quantities $\Lambda^l_{ijk;h}$ by

$\dfrac{\partial \phi^l_{ijk}}{\partial \Lambda^{(s)}} \Lambda^{(s)}_{;h}$. It is worth remarking that the quantities $\Lambda^{(s)}_{;h}$ are themselves 16 of the functions ϕ^l_{ijk}. With this substitution, equations (I) and (II) become a system of *ordinary* partial differential equations in the unknown functions ϕ^l_{ij} and ϕ^l_{ijk} of $\Lambda^{(1)}, \Lambda^{(2)}, \Lambda^{(3)}, \Lambda^{(4)}$. Moreover, one has

$$
\phi^l_{ijk} = \frac{\partial \phi^l_{ij}}{\partial \Lambda^{(s)}} \Lambda^{(s)}_{;k}.
$$

Finally, there are $24 - 4 = 20$ functions ϕ^l_{ij} and 96 functions ϕ^l_{ijk}.

It could easily be shown, though it is not essential, that this system of ordinary partial differential equations is in involution. The analysis made of the respective numbers of equations 1), 2), 3), 4) of system (II') show that if one orders the *new* equations (II), linear in

the $\dfrac{\partial \phi_{ijk}}{\partial \Lambda^{(s)}}$, with respect to the $\dfrac{\partial}{\partial \Lambda^{(1)}}, \dfrac{\partial}{\partial \Lambda^{(2)}}, \dfrac{\partial}{\partial \Lambda^{(3)}}, \dfrac{\partial}{\partial \Lambda^{(4)}}$ derivatives, one will arrive at the same numbers (the value of which is always the same, whatever the choice of the variables provided they are non-characteristic). As a consequence:

1° il existe dans le nouveau système (II), 96 équations linéairement indépendantes par rapport aux 96 dérivées $\dfrac{\partial \phi_{ijk}^l}{\partial \Lambda^{(4)}}$;

2° il n'existe que $96 - 12$ équations ne contenant aucune des dérivées $\dfrac{\partial}{\partial \Lambda^{(4)}}$ et linéairement indépendantes par rapport aux $\dfrac{\partial \phi_{ijk}^l}{\partial \Lambda^{(3)}}$.

De la première propriété on déduit que toute solution du système différentiel qui donne les ϕ_{ij}^k et ϕ_{ijh}^k est complètement déterminée par la solution à 3 dimensions dans la section $\Lambda^{(4)} = $ Cte.

De la seconde propriété il résulte que, dans cette section, on peut se donner arbitrairement 12 des fonctions inconnues ϕ_{ijh}^k, *mais qu'on ne peut pas s'en donner plus de 12.*

Par suite l'indice de généralité cherché est 12.

En dernière analyse la conclusion est fondée exclusivement sur la propriété du système (I) de contenir 24 équations ($C_k = 0$) linéairement indépendantes par rapport aux $\Lambda_{ij;4}^k$ et de n'en contenir que $24 - 12$ ($B_j = 0$) linéairement indépendantes par rapport aux $\Lambda_{ij;3}^k$ et ne contenant pas les $\Lambda_{ij;4}^k$. C'est ce *déficit* 12 qui se maintient quand on introduit les dérivées secondes et qui indique l'indice de généralité cherché. Ce nombre 12 peut donc en définitive s'obtenir en partant des équations du champ *mais en ne tenant pas compte des fonctions $h_{s\alpha}$.*

C'est donc en allant au fond des choses, c'est-à-dire en cherchant à reconnaître comment deux espaces riemanniens à parallélisme absolu sont identiques, qu'on peut trouver par une méthode sûre le degré de généralité *essentiel* de ces espaces.

Dans le cas du système $R_{ik} = 0$ qui fait intervenir les espaces riemanniens *sans* parallélisme absolu, la démonstration est plus compliquée, parce qu'on ne trouve pas aussi rapidement les conditions nécessaires et suffisantes d'application de deux espaces riemanniens. Ce sera si vous le voulez bien pour une prochaine lettre.

En tous cas ne vous apitoyez pas sur ma patience. Au fond je vous suis très reconnaissant de m'avoir obligé à chercher, pour des théorèmes que je possède depuis longtemps, une forme de présentation qui ne soit pas trop difficile à assimiler pour des mathématiciens qui n'ont pas de connaissances trop spéciales en Analyse. C'est donc un grand service que vous me rendez!

Veuillez agréer, cher et illustre Maître, l'expression de mes sentiments bien cordiaux et dévoués.

E. Cartan

1. there exist, in the new system (II), 96 equations linearly independent with respect to the 96 derivatives $\frac{\partial \phi^l_{ijk}}{\partial \Lambda^{(4)}}$;

2. there exist only $96 - 12$ equations containing no $\frac{\partial}{\partial \Lambda^{(4)}}$ derivatives that are linearly independent with respect to the $\frac{\partial \phi^l_{ijk}}{\partial \Lambda^{(3)}}$.

From the first property one deduces that any solution of the differential system that gives the ϕ^k_{ij} and ϕ^k_{ijh} is completely determined by the 3-dimensional solution en the section $\Lambda^{(4)} = $ const.

From the second property, it follows that, on this section, 12 of the unknown functions $\phi_{ijh}{}^k$ can be given arbitrarily *but no more than 12*. As a result, the generality index one is looking for is 12.

Observe that the conclusion relies exclusively on the fact that the system (I) contains 24 equations ($C_k = 0$) which are linearly independent with respect to the $\Lambda^k_{ij;4}$, and contains only $24 - 12$ ($B_j = 0$) equations which are linearly independent with respect to the $\Lambda^k_{ij;3}$ and do not contain the $\Lambda^k_{ij;4}$. It is this *deficit* of 12 that is maintained when one introduces the second derivatives and that indicates the generality index. This number 12 can thus be obtained from the field equations but only *by not taking the functions $h_{s\alpha}$ into account*.

So, it is only by delving deeply into things, that is by trying to recognize how two Riemannian spaces with absolute parallelism are identical, that one can find by a secure method, the *essential* degree of generality of these spaces in a sure way.

In the case of the system $R_{ik} = 0$, in Riemannian spaces *without* absolute parallelism, the proof is more complicated, since one cannot so quickly find the necessary and sufficient conditions for the applicability of two Riemannian spaces. If you wish this will be for another letter.

In any case, dont't feel sorry for my patience. After all, I am very grateful to you for having obliged me to find, in the case of theorems that I have known for a long time, a form of presentation that would not be too difficult for mathematicians having no special knowledge of analysis to understand. Thus, you have actually done me a great favor!...

E. Cartan

XXV

30-I-30

Verehrter Herr Cartan !

Ich muss nochmals auf den *Index de généralité* zurückkommen. Es scheint mir möglich, dass folgende Aenderung der Betrachtungsweise zweckmässig wäre. Man begnügt sich für die Charakterisierung nicht mit *einer* Zahl sondern geht es so vor:

Wie Sie es gethan haben, eliminiert man aus den n Gleichugen (wenn möglich) zuerst die $\frac{\partial f}{\partial x^4}$. Man erhält so r_4 Gleichungen, welche $\frac{\partial f}{\partial x^4}$ enthalten und $n - r_4$, welche die $\frac{\partial f}{\partial x^4}$ nicht enthalten. Hierauf eliminiert man aus den $n - r_4$ Gleichungen die $\frac{\partial f}{\partial x^3}$. Man erhält so r_3 Gleichungen, welche die $\frac{\partial f}{\partial x^3}$ $\left(\text{nicht aber die } \frac{\partial f}{\partial x^4}\right)$ enthalten, sowie $n - r_4 - r_3$ Gleichungen, welche weder die $\frac{\partial f}{\partial x_4}$ noch die $\frac{\partial f}{\partial x^3}$ enthalten. So zerlegt man in Ihrer Weise weiter, wobei man die Zahlen

$$r_4, r_3, r_2, r_1, \text{ (bei 4 Dimensionen)}$$

erhält, wobei $n = r_4 + r_3 + r_2 + r_1$.

Ist das System in Involution und p die Zahl der Variabeln f, so ist die einzelne Lösung des Systems durch folgende freie Angaben determiniert:

$$(\mathscr{I}_1 =)p - r_1 \text{ Funktionen von } x_1;$$
$$(\mathscr{I}_2 =)p - r_2 \text{ Funktionen von } x_1, x_2;$$
$$(\mathscr{I}_3 =)p - r_3 \text{ Funktionen von } x_1, x_2, x_3;$$
$$(\mathscr{I}_4 =)p - r_4 \text{ Funktionen von } x_1, x_2, x_3, x_4.$$

Dabei ist die freie Wahl der Funktionen nur durch die Bedingung eingeschränkt, dass die Fortsetzung der Lösung in das Kontinuum von nur 1 höherer Dimensionszahl jeweilen stetig sich anschliessen muss.

A XXV

30-I-30

Dear M. Cartan,

Once again I must return to the *indice de généralité*. It seems possible to me that the following change of approach might be appropriate. One doesn't stop with just *one* number for its characterization but proceeds as follows:

As you have done, one first eliminates the $\frac{\partial f}{\partial x^4}$ from the n equations (if possible). Thus one obtains r_4 equations which contain $\frac{\partial f}{\partial x^4}$ and $n - r_4$ which do not contain $\frac{\partial f}{\partial x^4}$. Then, from the $n - r_4$ equations, one eliminates $\frac{\partial f}{\partial x^3}$. Thus one obtains r_3 equations which contain $\frac{\partial f}{\partial x^3}$ $\left(\text{but not } \frac{\partial f}{\partial x^4} \right)$, and $n - r_4 - r_3$ equations which contain neither $\frac{\partial f}{\partial x_4}$ nor $\frac{\partial f}{\partial x^3}$. One continues decomposing in your way until one obtains the numbers

$$r_4, r_3, r_2, r_1, \text{ (for 4 dimensions)},$$

where $n = r_4 + r_3 + r_2 + r_1$.

If the system is in involution and p is the number of variables f, then the unique solution of the system is determined by the following independent specifications:

$(\mathscr{I}_1 =)p - r_1$ functions of x_1;
$(\mathscr{I}_2 =)p - r_2$ functions of x_1, x_2;
$(\mathscr{I}_3 =)p - r_3$ functions of x_1 , x_2, x_3;
$(\mathscr{I}_4 =)p - r_4$ functions of x_1, x_2, x_3, x_4.

Hence, the free choice of functions is restricted only by the condition that the continuation of the solution in the continuum of 1 higher dimension must join continuously onto that solution. For example, for the system of equations

Für das Gleichungssystem

$\Lambda^{\alpha}_{\mu\nu} = 0$ (Euklidisch mit gewöhnlichen Parallelismus) ist z.B.

$\mathscr{I}_4$	$\mathscr{I}_3$	$\mathscr{I}_2$	$\mathscr{I}_1$
4	8	12	16

Für mein Gleichungssystem dürfte sein

$\mathscr{I}_4$	$\mathscr{I}_3$	$\mathscr{I}_2$	$\mathscr{I}_1$
4	20	34	40

$\mathscr{I}_3 - (\mathscr{I}_3)_{eukl.}$ entspricht natürlich Ihren Index $\mathscr{I}$, mit Rücksicht auf die allgemeine Kovarianz.

Die einzige Frage bleibt nun die, *wie man den Grad der Determination am natürlichsten festlegt, wenn es sich darum handelt, den Determinations-Grad zweier Gleichungssysteme miteinander zu vergleichen.*

Soll man fragen:

,,Wieviel Funktionen können wir in einem Schnitt $x^4 =$ konst. frei wählen?" Oder soll man den Determinationsgrad durch die Reihe $\mathscr{I}_4, \mathscr{I}_3, \mathscr{I}_2, \mathscr{I}_1$ charakterisieren?

Die ganze Frage ist ja dadurch aufgetreten, weil wir feststellen wollten, ob das Gleichungssystem

$$R_{ik} = 0, \quad \Lambda^{\alpha}_{\mu\alpha} = \frac{1}{\psi}\frac{\partial\psi}{\partial x^{\mu}}, \quad S_{\mu} = \frac{1}{\psi}\frac{\partial\chi}{\partial x^{\mu}},$$

und das Gleichungssystem

$$\Lambda^{\alpha}_{\mu\nu;\nu} - \Lambda_{\mu}{}^{\sigma}{}_{\tau}\Lambda^{\alpha}_{\sigma\tau} = 0, \quad \Lambda^{\alpha}_{\mu\nu;\alpha} = 0,$$

punkto Determination einander gleichwertig seien.

Ich neige der Auffassung zu, dass die zweite Betrachtungsweise wegen ihrer Vollständigkeit zu bevorzugen sei. Wie denken Sie darüber?

Herzlich grüsst Sie

Ihr

A. Einstein

P.S. Ich las soeben Ihren sehr eigehenden Brief und bin — ohne die Einzelheiten schon erfassen zu können — von der Sicherheit der Ergebnisse *völlig überzeugt*. Ich sende diesen Brief trotzdem einstweilen ab, weil er durch dessen Inhalt nicht überflüssig gemacht wird.

$\Lambda^{\alpha}_{\mu\nu} = 0$ (Euclidean with ordinary parallelism)

$\mathscr{I}_4$	$\mathscr{I}_3$	$\mathscr{I}_2$	$\mathscr{I}_1$
4	8	12	16

For my system of equations we should have

$\mathscr{I}_4$	$\mathscr{I}_3$	$\mathscr{I}_2$	$\mathscr{I}_1$
4	20	34	40

Of course, $\mathscr{I}_3 - (\mathscr{I}_3)_{eucl.}$ corresponds to your index $\mathscr{I}$, taking general covariance into account.

The only question now remaining is, *how does one fix the degree of determination in the most natural way when it comes to comparing the degrees of determination of two systems of equations.*

Should one ask " How many functions can be freely chosen on a section $x^4 = $ const.? " or should one characterize the degree of determination by means of the series $\mathscr{I}_4$, $\mathscr{I}_3$, $\mathscr{I}_2$, $\mathscr{I}_1$?

The whole question has arisen because we would like to establish whether the system of equations

$$R_{ik} = 0, \quad \Lambda^{\alpha}_{\mu\alpha} = \frac{1}{\psi}\frac{\partial\psi}{\partial x^{\mu}}, \quad S_{\mu} = \frac{1}{\psi}\frac{\partial\chi}{\partial x^{\mu}},$$

and the system of equations

$$\Lambda^{\alpha}_{\mu\nu;\nu} - \Lambda^{\ \sigma}_{\mu\ \tau}\Lambda^{\alpha}_{\sigma\tau} = 0, \quad \Lambda^{\alpha}_{\mu\nu;\alpha} = 0,$$

are equivalent as regards degree of determination.

I incline to the opinion that the second approach is to be preferred because of its completeness. What do you think?

Kind regards.

Yours,

A. Einstein

P.S. I have just now read your very thorough letter and — without yet grasping all the details — I am *completely convinced* of the reliability of the results. Nevertheless, I am sending this letter because its contents have not been rendered completely superfluous.

XXVI

31-I-30

Verehrter Herr Cartan!

In meinem gestrigen Briefe war verschiedenes unrichtig, aber der Grundgedanke erscheint mir doch vernünftig, nämlich den Allgemeinheitsgrad durch die Zahlenfolge $\mathscr{I}_4$, $\mathscr{I}_3$, $\mathscr{I}_2$, $\mathscr{I}_1$ zu charakterisieren. Die Zahlenangaben sowie der Vergleich mit dem euklidischen Falle aber waren unrichtig.

Zur Bestimmung der $\mathscr{I}$ im Falle unserer Gleichungssysteme erscheint mir folgende Methode als einfach. Wir wählen als Variable die Grössen h_{sv} und $h_{sv\alpha}\left(= \dfrac{\partial h_{sv}}{\partial x^\alpha}\right)$. Ich betrachte nun mein System $G^{\mu\alpha} = 0$, $F^{\mu\alpha} = 0$.

Dies sind dann Gleichungen erster Ordnung in unseren Variabeln. Ich benutze die beiden Identitäten

$$\Lambda^\alpha_{\mu\nu;\alpha} \equiv \frac{\partial \phi_\mu}{\partial x^\nu} - \frac{\partial \phi_\nu}{\partial x^\mu}, \tag{1}$$

$$2\overline{G}^{\mu\alpha} - F^{\mu\alpha} \equiv S^\alpha_{\underline{\mu\nu};\nu} + S^\alpha_{\underline{\mu\sigma}}\Lambda^\nu_{\sigma\nu}. \tag{2}$$

Es können also 12 der 22 Gleichungen in der Form

$$0 = \frac{\partial \phi_\mu}{\partial x^\nu} - \frac{\partial \phi_\nu}{\partial x^\mu}, \tag{3}$$

$$0 = S^\alpha_{\underline{\mu\nu};\nu} + S^\alpha_{\underline{\mu\sigma}}\Lambda_\sigma{}^\nu_\nu, \tag{4}$$

geschrieben werden. 6 von diesen Gleichungen enthalten keine nach x^4 differenzierten Glieder, zwei davon auch keine nach x^3 differenzierten Glieder. Man hat also unmittelbar

10 + 6 Gleichungen, die Differentialquotienten nach allen Variabeln enthalten
4 Gleichungen, die Differentialquotienden nach x_1, x_2, x_3 enthalten
2 Gleichungen, die Differentialquotienten nach x_1, x_2 enthalten.

Wegen der allgemeinen Kovarianz können aus 4 der 16 Gleichungen die Ableitungen nach x^4 eliminiert werden, was man auch daraus

A XXVI

31.I.30

Dear M. Cartan,

Several things were wrong in yesterday's letter but the basic idea still seems reasonable to me, namely, to characterize the degree of generality by the series of numbers $\mathscr{I}_4$, $\mathscr{I}_3$, $\mathscr{I}_2$, $\mathscr{I}_1$. But the numbers given, and the comparison with the Euclidean case, were wrong.

The following method of fixing $\mathscr{I}$ in the case of our system of equations seems simple to me. We choose as variables the quantities h_{sv} and $h_{sv\alpha}\left(=\dfrac{\partial h_{sv}}{\partial x^\alpha}\right)$. I now look at my system, $G^{\mu\alpha} = 0$, $F^{\mu\alpha} = 0$. These are now equations of the first order in our variables. I use the two identities

$$\Lambda^\alpha_{\mu v;\alpha} \equiv \frac{\partial \phi_\mu}{\partial x^v} - \frac{\partial \phi_v}{\partial x^\mu}, \tag{1}$$

$$2\overline{G}^{\mu\alpha} - F^{\mu\alpha} \equiv S^\alpha_{\underline{\mu v};v} + S^\alpha_{\underline{\mu\sigma}}\Lambda^v_{\sigma v}. \tag{2}$$

Then 12 of the 22 equations can be written in the form

$$0 = \frac{\partial \phi_\mu}{\partial x^v} - \frac{\partial \phi_v}{\partial x^\mu}, \tag{3}$$

$$0 = S^\alpha_{\underline{\mu v};v} + S^\alpha_{\underline{\mu\sigma}}\Lambda_\sigma{}^v{}_v, \tag{4}$$

Of these equations, 6 contain no terms differentiated with respect to x^4, two of them having no derivatives with respect to x^3 either. Thus one has immediately:

10 + 6 equations which contain derivatives with respect to all variables,
4 equations which contain derivatives with respect to x_1, x_2, x_3,
2 equations which contain derivatives with respect to x_1, x_2.

Because of general covariance, derivatives with respect to x^4 can be eliminated from 4 of the 16 equations, something which can also be

ersieht, dass von den drei Vierer-Identitäten bisher nur 2 benutzt worden sind. Diese 4 durch die Elimination entstehenden Gleichungen enthalten die Indizes 1, 2 und 3. Die ursprünglichen Gleichungen lassen sich also so schreiben

12 mit den Differentiationen nach x_4, x_3, x_2, x_1,
8 mit den Differentiationen nach x_3, x_2, x_1,
2 mit den Differentiationen nach x_2, x_1.

Das vollständige Gleichungssystem lautet bei unserer Wahl der Variabeln

$$\left.\begin{array}{l} G^{\mu\alpha} = 0 \\ F^{\mu\alpha} = 0 \end{array}\right\} \quad\begin{array}{cccc} 12 & 8 & 2 & 0 \end{array}$$

$$h_{s\mu,\sigma} - h_{s\mu\sigma} = 0 \quad \begin{array}{cccc} 16 & 16 & 16 & 16 \end{array}$$

$$h_{s\mu\sigma,\tau} - h_{s\mu\tau,\sigma} = 0 \quad \begin{array}{cccc} 48 & 32 & 16 & 0 \end{array}$$

Die daneben geschriebenen Zahlen geben an, in wievielen der Gleichungen jeweilen nach wieviel Variabeln differenziert ist. Nun erhalten wir die Zahlen r_4, r_3, r_2, r_1 unseres Gleichungssystems

r_4	r_3	r_2	r_1
76	56	34	16

Die Zahl der Variabeln ist $p = 80$. Also erhalten wir für die $\mathscr{I}$

$\mathscr{I}_4$	$\mathscr{I}_3$	$\mathscr{I}_2$	$\mathscr{I}_1$
4	24	46	64

Nun vergleichen wir dies *sachgemäss* (nicht wie in meinem vorigen Brief!) mit dem euklidischen Falle, indem wir die h_{sv} und $h_{sv\alpha}$ als Variabeln nehmen. Wir erhalten dann in analoger Bezeichnungsweise

$$h_{s\mu,v} - h_{sv,\mu} = 0 \quad \begin{array}{cccc} 12 & 8 & 4 & 0 \end{array}$$

$$h_{s\mu,v} - h_{s\mu v} = 0 \quad \begin{array}{cccc} 16 & 16 & 16 & 16 \end{array}$$

$$h_{s\mu\sigma,\tau} - h_{s\mu\tau,\sigma} = 0 \quad \begin{array}{cccc} 48 & 32 & 16 & 0 \end{array}$$

r_4	r_3	r_2	r_1
76	56	36	16

$\mathscr{I}_4$	$\mathscr{I}_3$	$\mathscr{I}_2$	$\mathscr{I}_1$
4	24	44	64

162

seen from the fact that, so far, only 2 of the three 4-identities have been used. The 4 equations arising from this elimination contain the indices 1, 2 and 3. The original equations may, therefore, be written:

12 with differentiation with respect to $x_4, x_3, x_2, x_1,$
8 with differentiation with respect to $x_3, x_2, x_1,$
2 with differentiation with respect to $x_2, x_1.$

With our choice of variables, the complete system reads

$$
\begin{array}{ll}
\left.\begin{array}{l} G^{\mu\alpha} = 0 \\ F^{\mu\alpha} = 0 \end{array}\right\} & 12 \quad 8 \quad 2 \quad 0 \\
h_{s\mu,\sigma} - h_{s\mu\sigma} = 0 & 16 \quad 16 \quad 16 \quad 16 \\
h_{s\mu\sigma,\tau} - h_{s\mu\tau,\sigma} = 0 & 48 \quad 32 \quad 16 \quad 0.
\end{array}
$$

The numbers written alongside the equations state how many of the equations are differentiated with respect to how many of the variables. We now obtain the numbers r_4, r_3, r_2, r_1 for our system of equations:

r_4	r_3	r_2	r_1
76	56	34	16

The number of variables is $p = 80$. Hence, we obtain for the $\mathscr{I}$'s

$\mathscr{I}_4$	$\mathscr{I}_3$	$\mathscr{I}_2$	x_1
4	24	46	64

Now we compare this *properly* (and not as in my previous letter!) with the Euclidean case, in which we take h_{sv} and $h_{sv\alpha}$ as variables. We then obtain, in an analogous notation:

$$
\begin{array}{ll}
h_{s\mu,v} - h_{sv,\mu} = 0 & 12 \quad 8 \quad 4 \quad 0 \\
h_{s\mu,v} - h_{s\mu v} = 0 & 16 \quad 16 \quad 16 \quad 16 \\
h_{s\mu\sigma,\tau} - h_{s\mu\tau,\sigma} = 0 & 48 \quad 32 \quad 16 \quad 0
\end{array}
$$

r_4	r_3	r_2	r_1
76	56	36	16

$\mathscr{I}_4$	$\mathscr{I}_3$	$\mathscr{I}_2$	$\mathscr{I}_1$
4	24	44	64

Bei dem Vergleich gibt sich erst in der dritten Kolonne ($\mathscr{I}_2$) ein Unterscheid zwischen beiden Fallen! Die Determination der Mannigfaltigkeit scheint also eine geradezu unheimliche zu sein; es ist wahrhaft paradox. Finden Sie dass ich einen „ Bock geschossen " (d.h. falsch geschlossen) habe?

Herzlich grüsst Sie

Ihr

A. Einstein

Post-skriptum

1) Genau die gleichen Verhältnisse finde ich vor bei Ihrem Gleichungssystem

$$\mathrm{R}_{ik} = 0, \quad \frac{\partial \phi_i}{\partial x^k} - \frac{\partial \phi_k}{\partial x^i} = 0, \quad \frac{\partial \mathrm{S}^i}{\partial x^k} - \frac{\partial \mathrm{S}^k}{\partial x^i} = 0.$$

Auch ist dabei interessant, dass das zweite und dritte System noch Zusatz-Bedingungen für die g_{ik} liefert, sodass mein Einwand gegen dieses System nicht ganz berechtigt war. *Auch die Maxwell'schen Gleichungen kommen in erster Näherung heraus.* Es wird also notwendig sein, dies System in Betracht zu ziehen.

2) Auch die *rein* Riemann'sche Theorie lässt sich bei dieser Rechenart bequem mit Euklied vergleichen. Die Ergebnisse sind so:

$R_{ik} = 0$				$R_{ik,lm} = 0$			
$\mathscr{I}_4$	$\mathscr{I}_3$	$\mathscr{I}_2$	$\mathscr{I}_1$	$\mathscr{I}_4$	$\mathscr{I}_3$	$\mathscr{I}_2$	$\mathscr{I}_1$
4	16	30	40	4	12	24	40

Die rein Riemann'sche Relativitätstheorie beschränkt also nach diesen Ergebnissen schwächer als die neue Theorie.

Ich bin sehr gespannt darauf, ob Sie die Methode billigen und ob Sie in diesen Resultaten Fehler finden. Es ist ja wirklich kaum glaublich, dass in der neuen Theorie die Determination so gewaltig sein soll, zumal dies bei der ersten Näherung nicht hervortritt.

Comparing, a difference between the two cases appears only in the third column ($\mathscr{I}_2$)! Thus, the determination of the manifold appears to be almost uncanny; a true paradox. Do you think I have " shot a buck " (i.e. committed a blunder)?

Kind regards.

Yours,

A. Einstein

Postscript

1) I find exactly the same situation with your system of equations

$$R_{ik} = 0, \frac{\partial \phi_i}{\partial x^k} - \frac{\partial \phi_k}{\partial x^i} = 0, \frac{\partial S^i}{\partial x^k} - \frac{\partial S^k}{\partial x^i} = 0.$$

It is interesting, too, that the second and third system yield additional conditions for the g_{ik}, so my objection to this system was not totally justified. *Even Maxwell's equations come out, in the first approximation.* Thus, it will be necessary to take account of this system.

2) The *pure* Riemannian theory is also easily compared with Euclid by this method. The results are as follows:

	$R_{ik} = 0$				$R_{ik,lm} = 0$		
$\mathscr{I}_4$	$\mathscr{I}_3$	$\mathscr{I}_2$	$\mathscr{I}_1$	$\mathscr{I}_4$	$\mathscr{I}_3$	$\mathscr{I}_2$	$\mathscr{I}_1$
4	16	30	40	4	12	24	40

Thus, according to these results, the pure Riemannian theory of relativity is more weakly restricted than the new theory.

I am very eager to find out whether you approve of the method and if you can find any errors in these results. In all honesty, it is scarcely believable that the determination should be so strong in the new theory, the less so since it is not so in the first approximation.

XXVII

Le Chesnay (Seine et Oise)
27 av. de Montespan
le 2 février 1930

Cher et illustre Maître,

J'ai bien reçu vos deux lettres du 30 et du 31 janvier. Les entiers $\mathscr{I}_1$, $\mathscr{I}_2$, $\mathscr{I}_3$, $\mathscr{I}_4$ peuvent en effet être tous considérés; malheureusement il n'y a que le dernier de ces nombres qui soit différent de zéro qui ait une signification mathématique précise. Voici en effet ce qui se passe.

Je conserve vos notations: p fonctions inconnues, $r_1 + r_2 + r_3 + r_4$ équations linéaires du premier ordre. Posons

$$s_1 = p - r_1,$$
$$s_2 = p - r_2,$$
$$s_3 = p - r_3,$$
$$s_4 = p - r_4.$$

Substituons maintenant au système donné le système obtenu en adjoignant les dérivées partielles du premier ordre comme nouvelles fonctions inconnues (système *prolongé*). Les nouvelles valeurs des entiers s_i sont

$$s'_1 = s_1 + s_2 + s_3 + s_4,$$
$$s'_2 = \quad\;\; s_2 + s_3 + s_4,$$
$$s'_3 = \qquad\qquad s_3 + s_4,$$
$$s'_4 = \qquad\qquad\qquad s_4.$$

Si $s_4 > 0$, la nouvelle valeur de s_4 reste la même, le nombre des fonctions qu'on peut se donner arbitrairement en x^1, x^2, x^3, x^4 est toujours le même. Mais *la valeur de s_3 a augmenté*. Dans ce cas ($s_4 > 0$), l'entier s_3 n'a donc aucune signification précise, et à plus forte raison les entiers s_1, s_2.

Si au contraire $s_4 = 0$, l'entier s_3 conserve sa valeur pour tous les systèmes prolongés et il a vraiment une signification intrinsèque. Mais il n'en est plus de même de s_2 et de s_1 (à moins que $s_3 = 0$, auquel cas c'est l'entier s_2 qui donne le degré de généralité).

166

A XXVII

Le Chesnay (Seine et Oise)
27 av. de Montespan
2 February 1930

Cher et illustre Maître,

I have received your letters of January 30 and 31. The integer $\mathscr{I}_1, \mathscr{I}_2, \mathscr{I}_3, \mathscr{I}_4$ can, indeed, all be taken into account; unfortunately, it is only the last non-vanishing number that has a precise mathematical significance. Here is what is happening.

I keep your notations: p unknown functions, $r_1 + r_2 + r_3 + r_4$ first order linear equations. We write

$$
\begin{aligned}
s_1 &= p - r_1, \\
s_2 &= p - r_2, \\
s_3 &= p - r_3, \\
s_4 &= p - r_4.
\end{aligned}
$$

Let us now substitute for the given system, that system obtained by adding on the first order partial derivatives as new unknown functions (the *extended* system). The new values of the integers s_i are

$$
\begin{aligned}
s_1' &= s_1 + s_2 + s_3 + s_4, \\
s_2' &= s_2 + s_3 + s_4, \\
s_3' &= s_3 + s_4, \\
s_4' &= s_4.
\end{aligned}
$$

If $s_4 > 0$, the new value of s_4 remains the same and the number of functions of x_1, x_2, x_3, x_4 that can be given arbitrarily is still the same. *But the value of s_3 has increased.* In this case ($s_4 > 0$) the integer s_3 has no precise significance, let alone the integers s_1, s_2.

If, on the contrary $s_4 = 0$, the integer s_3 keeps its value for any extended system and it has a true intrinsic meaning. But this is not so for s_2 and s_1 (unless $s_3 = 0$, in which case, it is the integer s_2 which gives the degree of generality).

On peut préciser tout cela. L'idéal serait de démontrer que, si $s_4 = 0$ et $s_3 = \mathscr{I} > 0$, l'entier s_3 est le même pour deux systèmes différentiels *équivalents*, c'est-à-dire tels qu'on puisse établir une correspondance biunivoque et continue entre toute solution du premier et toute solution du second. Sous cette forme générale, le problème me semble actuellement inabordable. Mais on peut — ce que j'ai fait dans la théorie des groupes infinis — donner une définition plus restrictive, mais encore très large, de deux systèmes différentiels équivalents. Soit un système Σ à p fonctions inconnues $f_1, f_2, ..., f_p$; adjoignons à ce système de nouvelles équations contenant les fonctions $f_1, ..., f_p$ et de nouvelles fonctions inconnues $f_{p+1}, ..., f_{p+q}$. On obtient ainsi un nouveau système Σ'. Je dirai que Σ' est un prolongement holoédrique de Σ si toute solution de Σ fournit pour $f_1, ..., f_p$ des valeurs susceptibles d'être associées à des valeurs de $f_{p+1}, ..., f_{p+q}$ de manière à obtenir une solution de Σ'. Je dirai alors que deux systèmes Σ_1 et Σ_2 sont équivalents, s'il existe deux prolongements holoédriques Σ'_1 et Σ'_2 de Σ_1 et Σ_2 qui puissent être ramenés l'un à l'autre par un changement de variables et de fonctions inconnues (les deux systèmes Σ'_1 et Σ'_2 comportent par suite le même nombre de fonctions inconnues). Dans ces conditions on a le théorème suivant:

Si pour le premier système Σ_1 les entiers $s_1, s_2, ..., s_n$ sont tels que

$$s_n = s_{n-1} = ... = s_{v+1} = 0, \quad s_v > 0,$$

on aura, pour le second système,

$$s'_n = s'_{n-1} = ... = s'_{v+1} = 0, \quad s'_v = s_v.$$

Mais laissons de côté ces généralités. Si pour comparer différents systèmes relativistes, vous prenez $\mathscr{I}_4 = 4$, cela enlève toute signification précise à $\mathscr{I}_3$. Pour que l'indice de généralité $\mathscr{I}_3$ ait un sens, il faut donc considérer le système différentiel qui donne les propriétés *invariantes* des espaces, c'est-à-dire celui que je vous avais indiqué dans ma dernière lettre — à moins de particulariser les $h_{s\alpha}$ de manière à rendre $\mathscr{I}_4$ nul (et à donner à $\mathscr{I}_3$ sa vraie valeur).

J'ai été très intéressé par le post-scriptum de votre seconde lettre où vous ne rejetez pas absolument le système

$$R_{ik} = 0, \quad \frac{\partial \phi_i}{\partial x^k} - \frac{\partial \phi_k}{\partial x^i} = 0, \text{ etc...}$$

One can state all this more precisely. The best thing would be to prove that if $s_4 = 0$ and $s_3 = \mathscr{I} > 0$ then the integer s_3 is the same for two *equivalent* differential system, i.e. for systems such that a continuous one-to-one correspondence can be established between any solution of the first and any solution of the second. In this general form, the problem seems to me at present too difficult. But — and I have done this in the theory of infinite groups — one can give a more restrictive though still very broad definition of two equivalent differential systems. Let Σ be a system with p unknown functions $f_1, f_2, ..., f_p$; let us add to this system new equations containing the functions $f_1, ..., f_p$ and new unknown functions $f_{p+1}, ..., f_{p+q}$. One finds then a new system Σ'. I shall say that Σ' is a holohedral extension of Σ if any solution of Σ gives values for $f_1, ..., f_p$, that can be associated with values of $f_{p+1}, ..., f_{p+q}$ in such a way that one obtains a solution of Σ'. I shall also say that two systems Σ_1 and Σ_2 are equivalent of there exist two holohedral extensions Σ_1' and Σ_2' of Σ_1 and Σ_2 that can be reduced to each other by a change of variables and unknown functions (the two systems Σ_1' and Σ_2' have, therefore, the same number of unknown functions).

Under these circumstances one has the following theorem:

If for the first system Σ_1 the integers $s_1, s_2, ..., s_n$ are such that

$$s_n = s_{n-1} = s_{v+1} = 0, \quad s_v > 0,$$

one will have for the second system

$$s_n' = s_{n-1}' = ... = s_{v+1}' = 0, \quad s_v' = s_v.$$

But let us put aside these generalities. If in order to compare different relativistic systems, you take $\mathscr{I}_4 = 4$, this denies any precise meaning from $\mathscr{I}_3$. To give a meaning to the generality index $\mathscr{I}_3$ one has to consider the differential system that gives *invariant* properties of the spaces; in other words that one that I discussed in my last letter — unless one specializes the $h_{s\alpha}$ in such a way as to make $\mathscr{I}_4$ zero (and to give to $\mathscr{I}_3$ its true value).

I was very much interested in the postscript of your second letter where you don't absolutely reject the system

$$R_{ik} = 0, \quad \frac{\partial \phi_i}{\partial x^k} - \frac{\partial \phi_k}{\partial x^i} = 0, \text{ etc...}$$

169

etc. Vous êtez-vous amusé à chercher l'expression de $R_{\alpha\beta}$ au moyen des $\Lambda^{\gamma}_{\alpha\beta}$? On a

$$2R_{\alpha\beta} = \Lambda^{\beta}_{\alpha\mu;\mu} + \Lambda^{\alpha}_{\overline{\beta\mu};\mu} - \phi_{\alpha;\beta} - \phi_{\beta;\alpha} + \Lambda^{\rho}_{\alpha\mu}\Lambda^{\beta}_{\mu\rho} + \Lambda^{\rho}_{\beta\mu}\Lambda^{\alpha}_{\mu\rho} + S_{\alpha}S_{\beta}$$
$$- (\Lambda^{\beta}_{\overline{\alpha}\rho} + \Lambda^{\alpha}_{\overline{\beta}\rho})\phi_{\rho} + 2\Lambda^{\sigma}_{\alpha\rho}\Lambda^{\sigma}_{\overline{\beta}\rho} - g_{\alpha\beta}S_{\mu}S_{\underline{\mu}}.$$

Veuillez agréer, cher et illustre Maître, l'expression de mes sentiments tout dévoués.

E. Cartan

Je trouve ce matin dans Le Populaire une photographie, très ressemblante, vous représentant tenant votre violon à la sortie d'un concert [1]!

---

1. Le numéro du 2 février 1930 du journal français *Le Populaire* donne en effet en première page une photo d'Einstein tenant un violon, sous le titre: *Savant et Violoniste*, avec en légende: « Une bien curieuse nouvelle nous arrive de Berlin. Le professeur

etc. Would you like to know how to express $R_{\alpha\beta}$ in terms of $\Lambda^\gamma_{\alpha\beta}$? One has

$$2R_{\alpha\beta} = \Lambda^\beta_{\alpha\mu;\mu} + \Lambda^\alpha_{\beta\mu;\mu} - \phi_{\alpha;\beta} - \phi_{\beta;\alpha} + \Lambda^\rho_{\alpha\mu}\Lambda^\beta_{\mu\rho} + \Lambda^\rho_{\beta\mu}\Lambda^\alpha_{\mu\rho} + S_\alpha S_\beta$$
$$- (\Lambda^\beta_{\alpha\rho} + \Lambda^\alpha_{\beta\rho})\phi_\rho + 2\Lambda^\sigma_{\alpha\rho}\Lambda^\alpha_{\beta\rho} - g_{\alpha\beta}S_\mu S_\mu.$$

E. Cartan

This morning, in *Le Populaire,* I found a very good photograph of you, showing you coming out of a concert holding your violin [1]!

physicien Albert Einstein a joué, en qualité de violon solo, dans un grand concert de musique religieuse et classique. Voici l'illustre mathématicien à l'issue du concert. » Le 29 janvier 1930, Einstein jouait avec le prof. Lewandowski le largo du double concerto pour violon de J.-S. Bach au cours d'un concert donné dans le nouvelle synagogue de l'Oranienburgerstrasse à Berlin.

171

XXVIII

2-II-30

Verehrter Herr Cartan!

Nun habe ich selber den Fehler gefunden, der mein paradoxes Resultat verschuldet hat. Ich hatte geglaubt

$$\Lambda^\alpha_{\mu\nu} = 0 = h_{s\mu,\nu} - h_{s\nu,\mu},$$

die Gleichungen zufügen zu können

$$h_{s\mu,\nu} - h_{s\mu\nu} = 0,$$

sowie ein drittes System, das Ganze, damit das System formal gleichartig werde mit einem Gleichungssystem zweiter Ordnung. Ich hatte sonderbarer-weise übersehen, dass man mithilfe der zweiten Gleichung aus der ersten alle Ableitungen eliminieren kann! Der Vergleich meines Systems mit dem euklidischen war also falsch. Sonst aber finde ich die benutzte Form sehr einfach. Ich will den letzten Teil der Überlegung in verbesserter Form angeben.

Durch Betrachtungen wie sie in meinem Briefe angedeutet sind, gelangt man durch blosses Ordnen der Gleichungen zu folgenden drei Ergebnissen

	$\mathscr{S}_4$	$\mathscr{S}_3$	$\mathscr{S}_2$	$\mathscr{S}_1$
$R_{k\,lm}^{\,i} = 0$	4	12	24	40
$R_{ik} = 0$	4	16	30	40
Mein System	4	24	46	64

(Die Zahl der Identitäten ist immer $r_3 + 2r_2 + 3r_1$)

Aus der ersten Zeile sieht man, dass man in einem euklidischen Raume 12 der $g_{\mu\nu}$ und $g_{\mu\nu\alpha}$ in einem Raum von 3 Dimensionen frei wählen kann, durch passende Wahl des Koordinatensystems. Dies wird aber für jeden gegebenen Riemann-Raum gelten. Hat man ferner einen Raum mit Riemann-Metrik und Fernparallelismus, so kann

172

A XXVIII

2.II.30

Dear M. Cartan,

I have now myself found the error which is to blame for my paradoxical result. I thought that

$$\Lambda^{\alpha}_{\mu\nu} = 0 = h_{s\mu,\nu} - h_{s\nu,\mu},$$

were the equations to be added to

$$h_{s\mu,\nu} - h_{s\mu\nu} = 0,$$

in order to form a third system, consisting of all of them, which would then be formally analogous to a second order system of equations. Strangely enough, I had overlooked the fact that one can eliminate all the derivatives in the first equation by using the second! The comparison of my system with the Euclidean one is, therefore, wrong. But I can find the needed form very simply. I will give the last part of the analysis in an improved manner.

By means of considerations which I indicated to you in my letter, a simple ordering of the equations leads to the following three results

	$\mathscr{I}_4$	$\mathscr{I}_3$	$\mathscr{I}_2$	$\mathscr{I}_1$
$R_k{}^i{}_{lm} = 0$	4	12	24	40
$R_{ik} = 0$	4	16	30	40
My system	4	24	46	64

(The number of identities is always $r_3 + 2r_2 + 3r_1$).

One can see from the first row that, in a Euclidean space, 12 of the $g_{\mu\nu}$ and $g_{\mu\nu\alpha}$ can be freely chosen in a space of 3 dimensions by a suitable choice of coordinate system. Indeed this will hold for any given Riemannian space. Furthermore, if one has a space with a Riemannian metric and distant parallelism then, instead of the variables $h_{s\nu}$ one can choose the $g_{\mu\nu}$ and 6 further variables (e.g. 6 of the $h_{s\nu}$),

man statt der Variabeln h_{sv} die $g_{\mu v}$ und 6 weitere Variabeln (z.B. 6 der h_{sv}) nebst ersten Ableitungen dieser Grössen als Feld-variable wählen (und deren erste Ableitungen). Da sich 12 von den $g_{\mu v}$, $g_{\mu v\alpha}$ durch Koordinatenwahl in einer „ Fläche v. 3 Dim. " frei wählen lassen, so folgt, dass auch im h_{sv}-Raum 12 Variable frei wählbar sind auf einer „ Fläche " (durch Koordinatenwahl).

Es folgt also, dass abgesehen von der Koordinatenwahl

$$\text{im Falle } R_{ik} = 0, \ 16 - 12 = 4$$
$$\text{im Falle meiner Gleichungen, } 24 - 12 = 12$$

Feldvariable (abgesehen von der Koordinatenwahl) vorgeschrieben werden dürfen.

Die ist genau übereinstimmend mir Ihren Ergebnissen.

Eine Bemerkung möchte ich noch beifügen. Die Gleichungen $G^{\mu\alpha} = 0$ allein bilden schon ein Involutions-System wegen der Identität $G^{\mu\alpha}{}_{;\mu} + \Lambda^\alpha_{\sigma\tau} G^{\sigma\tau} \equiv 0$.

Daraus ensteht wieder ein Involutions-System durch Zufügen der Gleichungen $F^{\mu\alpha} = 0$. Ist es sicher, dass es nicht ein Involutions-System (zweiter Ordnung) gibt, welches durch Zufügen noch weiterer Gleichungen zu meinem System von Gleichungen entstehende? Es kommt mit nämlich so vor, als ob der *Index de généralité* meines Gleichungs-Systems noch zu hoch wäre. Dies wäre natürlich von höchster Wichtigkeit.

Herzlich grüsst Sie

Ihr

A. Einstein

174

together with their first derivatives, as field variables (and their first derivatives). Since 12 of the $g_{\mu\nu}$, $g_{\mu\nu\alpha}$ can be freely selected by a choice of coordinates on a « 3-dimensional surface », then in a h_{sv}-space, 12 variables are also freely specifiable (by means of a choice of coordinates) on a « surface ».

Thus it follows that, apart from a choice of coordinates,

$$\text{in the case of } R_{ik} = 0, \ 16 - 12 = 4$$
$$\text{in the case of my equations, } 24 - 12 = 12$$

field variables (apart from the choice of coordinates) may be prescribed.

This is in exact agreement with your results.

I would like to add one remark. The equations $G^{\mu\alpha} = 0$ alone form a system in involution because of the identity $G^{\mu\alpha}{}_{;\mu} + \Lambda^\alpha_{\sigma\tau}\, G^{\sigma\tau} \equiv 0$.

An involutive system again arises from adding on the equations $F^{\mu\alpha} = 0$. Is it certain that there exists no (second order) system in involution which might arise from the addition of still further equations to my system of equations? In particular, it seems to me that the *indice de généralité* of my system of equations might still be too high. Naturally, this would be of the greatest importance.

Kind regards.

Yours,

A. Einstein

XXIX

4-II-30

Verehrter Heer Cartan!

Ich muss Ihnen nochmals meine Bezeichnung erklären; ich glaube nämlich dass Sie mich bezüglich der $\mathscr{I}_4$, $\mathscr{I}_3$, $\mathscr{I}_2$, $\mathscr{I}_1$ missverstanden haben. Zuerst aber habe ich Furcht, nicht verstanden zu haben, was Sie unter „ *système prolongé* " verstehen; ist es so gemeint

$$a_{ikl}\frac{\partial f_k}{\partial x^l} + b_{ik}f_k = 0 \quad \text{(ursprüngliches System)}$$

$$\left.\begin{array}{l} a_{ikl}\phi_{kl} + b_{ik}f_k = 0, \\[2mm] \dfrac{\partial f_k}{\partial x^l} - \phi_{kl} = 0. \end{array}\right\} \quad \text{(système prolongé)}$$

Dies zweite System würde nicht unter das Schema fallen, weil es Gleichungen enthält, die überhaupt keine Ableitungen der unbekannten Funktionen enthalten. Aus diesen Gleichungen wären dann eben jene Unbekannten zu eliminieren.

Meine Bezeichnung war so: n Differentialgleichungen (algebraisch unabhängig) sind gegeben. Man eliminiert aus möglichst vielen von ihnen alle $\frac{\partial f}{\partial x^4}$, so dass man zunächst hat

r^4 Gleichungen, welche $\frac{\partial f}{\partial x^4}$ enthalten,

$n - r^4$ Gleichungen, welche $\frac{\partial f}{\partial x^4}$ nicht enthalten.

Nun eliminiert man die $\frac{\partial f}{\partial x^3}$ aus den $n - r^4$ Gleichungen. Es bleiben dann r^3 Gleichungen übrig, welche $\frac{\partial f}{\partial x^3}$ enthalten, und $n - r^4 - r^3$, welche weder die $\frac{\partial f}{\partial x^4}$ noch $\frac{\partial f}{\partial x^3}$ enthalen. So führt man fort und erhält das Quadrupel

$$r^4 \quad r^3 \quad r^2 \quad r^1$$

wobei

$$n = r^4 + r^3 + r^2 + r^1.$$

A XXIX

4.II.30

Dear M. Cartan,

I ought to explain my notation again; for I believe you have misunderstood me regarding the $\mathscr{I}_4, \mathscr{I}_3, \mathscr{I}_2, \mathscr{I}_1$. But first, I'm afraid I've not understood what you mean by a " *système prolongé* "; do you mean

$$a_{ikl} \frac{\partial f_k}{\partial x^l} + b_{ik} f_k = 0 \quad \text{(original system)}$$

$$\left. \begin{aligned} a_{ikl}\, \phi_{kl} + b_{ik} f_k &= 0, \\ \frac{\partial f_k}{\partial x^l} - \phi_{kl} &= 0. \end{aligned} \right\} \text{(système prolongé)}$$

The second system would not fit into the scheme since it contains equations which contain no derivatives of the unknown functions. Every single unknown would then be eliminated from such equations.

My notation was as follows: n (algebraically independent) differential equations are given. One eliminates all the $\frac{\partial f}{\partial x^4}$, from as many of them as possible, and one ends up with

$$r^4 \text{ equations which contain } \frac{\partial f}{\partial x^4},$$
$$n - r^4 \text{ equations which do not contain } \frac{\partial f}{\partial x^4}.$$

Now one eliminates the $\frac{\partial f}{\partial x^3}$ from the $n - r^4$ equations. Then there remain r^3 equations which contain the $\frac{\partial f}{\partial x^3}$ and $n - r^4 - r^3$ which contain neither the $\frac{\partial f}{\partial x^4}$ nor the $\frac{\partial f}{\partial x^3}$. One continues in this way and, finally, obtains the quadruple

$$r^4 \quad r^3 \quad r^2 \quad r^1$$

where

$$n = r^4 + r^3 + r^2 + r^1.$$

177

Damit das System in Involution sei, müssen $r^3 + 2r^2 + 3r^1 \left(\sum\limits_{v} v \, r^{d-v} \right)$ Identitäten von der von Ihnen angegeben Art bestehen.

Setzt man $\mathscr{I}^4 = p - r^4$ (p = Zahl der Variabeln f) etc. so charakterisiert $\mathscr{I}^4, \mathscr{I}^3, \mathscr{I}^2, \mathscr{I}^1$ den Determinations-Charakter des Systems vollständig, indem in einem (Teil) Kontinuum von α Dimensionen $\mathscr{I}^\alpha$ Funktionen frei wählbar sind.

Beispiel $\dfrac{\partial f_\alpha}{\partial x^\beta} - \dfrac{\partial f_\beta}{\partial x^\alpha} = 0$.

Wir haben als Ergebnis des Eliminationsverfahrens

			Zahl der Identitäten
$\dfrac{\partial f_1}{\partial x^4} - \dfrac{\partial f_4}{\partial x^1} = 0$	$\dfrac{\partial f_1}{\partial x^3} - \dfrac{\partial f_3}{\partial x^1} = 0$	$\dfrac{\partial f_1}{\partial x^2} - \dfrac{\partial f_2}{\partial x^1} = 0$	$r^3 + 2r^2 = 4$
$\dfrac{\partial f_2}{\partial x^4} - \dfrac{\partial f_4}{\partial x^2} = 0$	$\dfrac{\partial f_2}{\partial x^3} - \dfrac{\partial f_3}{\partial x^2} = 0$		dies stimmt ebenfalls
$\dfrac{\partial f_3}{\partial x^4} - \dfrac{\partial f_4}{\partial x^3} = 0$			$\dfrac{\partial}{\partial x^\gamma} \left(\dfrac{\partial f_\alpha}{\partial x^\beta} - \dfrac{\partial f_\beta}{\partial x^\alpha} \right) +$ ... + ... $\equiv 0$.
$r^4 = 3$	$r^3 = 2$	$r^2 = 1$	$r^1 = 0$
$\mathscr{I}^4 = 1$	$\mathscr{I}^3 = 2$	$\mathscr{I}^2 = 1$	$\mathscr{I}^1 = 4$.

Das Gesagte soll natürlich gegenüber Ihrer schönen Darstellung gar nichts Neues sein sondern nur eine übersichtliche Darstellung des Resultates.

Was den Zusammenhang Ihres Gleichungssystems mit dem meinigen betrifft, so habe ich folgendes gesehen. Es ist gesetzt

$$\Lambda^\sigma_{\mu\sigma} = \phi_\mu = \frac{\partial \log \psi}{\partial x^\mu}.$$

Die Mannigfaltigkeit hat auch eine Metrik

$$g_{\mu v} = \psi \, h_{s\mu} \, h_{sv}.$$

Bildet man $R^{\mu v}$ aus *diesen* $g_{\mu v}$, so enthält

$$2R^{\mu\alpha} - (G^{\mu\alpha} + G^{\alpha\mu})$$

jedenfalls keine zweimal differenzierten h mehr. Es scheint aber nach dem von Ihnen angegebenen Ausdruck, dass doch quadratische Glie-

In order that the system be in involution, $r^3 + 2r^2 + 3r^1$ $(\sum_v v\, r^{d-v})$, identities of the kind given by you must hold.

By putting $\mathscr{I}^4 = p - r^4$ (p = number of variables $_J$), etc., $\mathscr{I}^4$ $\mathscr{I}^3$, $\mathscr{I}^2$, $\mathscr{I}^1$, completely characterize the nature of the determination of the system, in which, in a (part of a) continuum of α dimensions, $\mathscr{I}^\alpha$ functions are freely specifiable.

An example: $\dfrac{\partial f_\alpha}{\partial x^\beta} - \dfrac{\partial f_\beta}{\partial x^\alpha} = 0.$

We have, as a result of the elimination procedure,

			Number of identities
$\dfrac{\partial f_1}{\partial x^4} - \dfrac{\partial f_4}{\partial x^1} = 0$	$\dfrac{\partial f_1}{\partial x^3} - \dfrac{\partial f_3}{\partial x^1} = 0$	$\dfrac{\partial f_1}{\partial x^2} - \dfrac{\partial f_2}{\partial x^1} = 0$	$r^3 + r^2 = 4$
$\dfrac{\partial f_2}{\partial x^4} - \dfrac{\partial f_4}{\partial x^2} = 0$	$\dfrac{\partial f_2}{\partial x^3} - \dfrac{\partial f_3}{\partial x^2} = 0$		this agrees with
$\dfrac{\partial f_3}{\partial x^4} - \dfrac{\partial f_4}{\partial x^3} = 0$			$\dfrac{\partial}{\partial x^\gamma}\left(\dfrac{\partial f_\alpha}{\partial x^\beta} - \dfrac{\partial f_\beta}{\partial x^\alpha}\right) +$ $\dots + \dots \equiv 0.$

$r^4 = 3$	$r^3 = 2$	$r^2 = 1$	$r^1 = 0$
$\mathscr{I}^4 = 1$	$\mathscr{I}^3 = 2$	$\mathscr{I}^2 = 3$	$\mathscr{I}^1 = 4\,.$

Of course, what I have said is not at all new after your beautiful presentation, but merely a synoptic presentation of results.

As to the connexion of your system of equations with mine, I see it as follows. Let us set

$$\Lambda^\sigma_{\mu\sigma} = \phi_\mu = \frac{\partial \log \psi}{\partial x^\mu}.$$

The manifold also possesses a metric

$$g_{\mu\nu} = \psi h_{s\mu} h_{s\nu}.$$

If we construct $R^{\mu\nu}$ from *these* $g_{\mu\nu}$, then

$$2R^{\mu\alpha} - (G^{\mu\alpha} + G^{\alpha\mu})$$

in any case contains no more second derivatives of h. But it appears from your expression that quadratic terms still remain. For the special

der übrig bleiben. Denn für den Spezialfall $\psi = $ konst. erhält man aus der von Ihnen mitgeteilten Identität

$$2R_{\alpha\beta} - (G_{\alpha\beta} + G_{\alpha\beta}) \equiv 2\Lambda^{\sigma}_{\alpha\rho}\Lambda^{\sigma}_{\underline{\beta}\underline{\rho}}.$$

Die rechte Seite aber verschwindet *nicht* bei konstantem ψ. Es scheint also, dass sich beide Systeme nicht nur der Form sondern nach dem Inhalte nach voneinander unterscheiden. Dies scheint auch aus dem zentralsymmetrischen Spezialfall hervorzugehen.

Herzlich grüsst Sie Ihr

A. Einstein

case $\psi = const.$ using the identity which you sent to me, we obtain

$$2R_{\alpha\beta} - (G_{\alpha\beta} + G_{\alpha\beta}) \equiv 2\Lambda^\sigma_{\alpha\rho}\Lambda^\sigma_{\beta\underline{\rho}}.$$

But the right hand side does *not* vanish for constant ψ. So it appears that the two systems differ from each other in content as well as form. This also comes out in the spherically symmetric case.

Kind regards.

Yours,

A. Einstein

XXX

Le Chesnay (Seine et Oise)
27 avenue de Montespan,
le 7 février 1930

Cher et illustre Maître,

J'avais bien compris votre introduction des nouveaux entiers $\mathscr{I}_1$, $\mathscr{I}_2$, etc., que dans votre dernière lettre vous expliquez du reste très clairement. J'avais moi-même dans mes mémoires indiqué ces quantités que je désignais par s_1, s_2, etc. Ce que je voulais vous dire, c'est que un seul de ces entiers a une signification vraiment intrinsèque et que les autres dépendent de la forme analytique qu'on donne au système différentiel considéré. Voici ce que je veux dire.

Prenons le système que vous donnez comme exemple. Quand on construit la solution générale de ce système par la méthode de récurrence indiquée dans ma théorie des systèmes en involution, cette solution dépend bien effectivement de $\mathscr{I}_1$ fonctions arbitraires d'une variable, de $\mathscr{I}_2$ fonctions arbitraires de 2 variables, de $\mathscr{I}_3$ fonctions arbitraires de trois variables, etc. Mais d'abord, dans votre exemple,

$$\frac{\partial f_\alpha}{\partial x_\beta} - \frac{\partial f_\beta}{\partial x_\alpha} = 0,$$

il est bien clair pour quelqu'un qui n'est pas savant en mathématiques, que la solution générale dépend d'une fonction arbitraire ϕ de 4 variables:

$$f_\alpha = \frac{\partial \phi}{\partial x_\alpha},$$

et c'est tout: quand cette fonction est donnée, les fonctions inconnues sont parfaitement déterminées et on ne trouve plus trace des $\mathscr{I}_2 = 3$ fonctions arbitraires de 2 variables, etc. qui étaient nécessaires dans la méthode générale d'intégration.

Mais on peut aussi démontrer que la solution générale dépend de *plus* de fonctions arbitraires de 1, 2 et 3 variables que ne l'indiquent

182

A XXX

Le Chesnay (Seine et Oise)
27 av. de Montespan,
7 February 1930

Cher et illustre Maître,

I quite understand your introduction of the new integers $\mathscr{I}_1$, $\mathscr{I}_2$, etc. In any case, you explained these very clearly in your last letter. In my articles, I myself pointed out these quantities that I called s_1, s_2, etc. What I wanted to say to you is that only one of these integers has a true intrinsic value, and that the others depend on the analytic form of the differential system. Here is what I mean.

Let us take as an example the system that you give. When one constructs the general solution of this system by the recurrence method discussed in my theory of systems in involution, this solution turns out to depend on $\mathscr{I}_1$ arbitrary functions of one variable, $\mathscr{I}_2$ arbitrary functions of 2 variables, $\mathscr{I}_3$ arbitrary functions of three variables, etc. But first of all, in your example,

$$\frac{\partial f_\alpha}{\partial x_\beta} - \frac{\partial f_\beta}{\partial x_\alpha} = 0,$$

it is quite clear to somebody who is not well versed in mathematics that the general solution depends on an arbitrary function ϕ of 4 variables:

$$f_\alpha = \frac{\partial \phi}{\partial x_\alpha},$$

and that is all. When this function is given, the unknown functions are perfectly determined, and no trace is found of the $\mathscr{I}_2 = 3$ arbitrary functions of 2 variables, etc. that we needed in the general method of integration.

But it can also be shown that the general solution depends on *more* arbitrary functions of 1, 2, and 3 variables than is indicated by

183

les nombres $\mathscr{I}_1$, $\mathscr{I}_2$, $\mathscr{I}_3$. En effet *prolongeons* le système, c'est-à-dire écrivons-le sous la forme suivante

$$\frac{\partial f_\alpha}{\partial x_\beta} - f_{\alpha\beta} = 0, \ (f_{\alpha\beta} \equiv f_{\beta\alpha})$$

$$\frac{\partial f_{\alpha\beta}}{\partial x_\gamma} - \frac{\partial f_{\alpha\gamma}}{\partial x_\beta} = 0,$$

qui conduit naturellement pour les f_α à la même solution $f_\alpha = \dfrac{\partial \phi}{\partial x_\alpha}$, les $f_{\alpha\beta} \equiv f_{\beta\alpha}$ étant $\dfrac{\partial^2 \phi}{\partial x_\alpha \partial x_\beta}$. Écrivons ces équations par ordre. Nous obtenons

$$\frac{\partial f_\alpha}{\partial x_4} - f_{\alpha 4} = 0 \ (\alpha = 1,2,3,4) \qquad\qquad \frac{\partial f_\alpha}{\partial x_3} - f_{\alpha 3} = 0 \ (\alpha = 1,2,3,4)$$

$$\frac{\partial f_{1\alpha}}{\partial x_4} - \frac{\partial f_{4\alpha}}{\partial x_1} = 0 \ (\alpha = 1,2,3,4) \qquad \frac{\partial f_{1\alpha}}{\partial x_3} - \frac{\partial f_{3\alpha}}{\partial x_1} = 0 \ (\alpha = 1,2,3,4)$$

$$\frac{\partial f_{2\alpha}}{\partial x_4} - \frac{\partial f_{4\alpha}}{\partial x_2} = 0 \ (\alpha = 2,3,4)^* \qquad \frac{\partial f_{2\alpha}}{\partial x_3} - \frac{\partial f_{3\alpha}}{\partial x_2} = 0 \ (\alpha = 2,3,4)$$

$$\frac{\partial f_{3\alpha}}{\partial x_4} - \frac{\partial f_{4\alpha}}{\partial x_3} = 0 \ (\alpha = 3,4)$$

$$\frac{\partial f_\alpha}{\partial x_2} - f_{\alpha 2} = 0 \ (\alpha = 1,2,3,4) \qquad\qquad \frac{\partial f_\alpha}{\partial x_1} - f_{1\alpha} = 0 \ (\alpha = 1,2,3,4)$$

$$\frac{\partial f_{1\alpha}}{\partial x_2} - \frac{\partial f_{2\alpha}}{\partial x_1} = 0 \ (\alpha = 1,2,3,4)$$

* Je ne donne pas à α la valeur 1 dans la ligne 3 parce que les équations ne seraient pas toutes indépendantes. De même pour $\alpha = 2$ dans la ligne 4.

$$r_4 = 4 + 4 + 3 + 2 = 13, \quad r_3 = 4 + 4 + 3 = 11,$$
$$r_2 = 4 + 4 = 8, \quad r_1 = 4.$$

D'autre part, le nombre des fonctions inconnues est

$$4 \ (\text{pour les } f_\alpha) + 10 = 14,$$
$$\mathscr{I}_4 = 14 - r_4 = 1, \quad \mathscr{I}_3 = 14 - 11 = 3,$$
$$\mathscr{I}_2 = 14 - 8 = 6, \quad \mathscr{I}_1 = 14 - 4 = 10,$$

au lieu de

$$\mathscr{I}_4 = 1, \quad \mathscr{I}_3 = 2, \quad \mathscr{I}_2 = 3, \quad \mathscr{I}_1 = 4.$$

184

the numbers $\mathscr{I}_1$, $\mathscr{I}_2$, $\mathscr{I}_3$. For let us extend the system, that is, write it in the following form

$$\frac{\partial f_\alpha}{\partial x_\beta} - f_{\alpha\beta} = 0, \ (f_{\alpha\beta} \equiv f_{\beta\alpha})$$

$$\frac{\partial f_{\alpha\beta}}{\partial x_\gamma} - \frac{\partial f_{\alpha\gamma}}{\partial x_\beta} = 0,$$

which naturally leads to the same solution $f_\alpha = \dfrac{\partial \phi}{\partial x_\alpha}$ for the f_α, the

$f_{\alpha\beta} \equiv f_{\beta\alpha}$ representing $\dfrac{\partial^2 \phi}{\partial x_\alpha \partial x_\beta}$. Let us order these equations. We get:

$$\frac{\partial f_\alpha}{\partial x_4} - f_{\alpha 4} = 0 \ (\alpha = 1,2,3,4) \qquad\qquad \frac{\partial f_\alpha}{\partial x_3} - f_{\alpha 3} = 0 \ (\alpha = 1,2,3,4)$$

$$\frac{\partial f_{1\alpha}}{\partial x_4} - \frac{\partial f_{4\alpha}}{\partial x_1} = 0 \ (\alpha = 1,2,3,4) \qquad \frac{\partial f_{1\alpha}}{\partial x_3} - \frac{\partial f_{3\alpha}}{\partial x_1} = 0 \ (\alpha = 1,2,3,4)$$

$$\frac{\partial f_{2\alpha}}{\partial x_4} - \frac{\partial f_{4\alpha}}{\partial x_2} = 0 \ (\alpha = 2,3,4)^* \qquad \frac{\partial f_{2\alpha}}{\partial x_3} - \frac{\partial f_{3\alpha}}{\partial x_2} = 0 \ (\alpha = 2,3,4)$$

$$\frac{\partial f_{3\alpha}}{\partial x_4} - \frac{\partial f_{4\alpha}}{\partial x_3} = 0 \ (\alpha = 3,4)$$

$$\frac{\partial f_\alpha}{\partial x_2} - f_{\alpha 2} = 0 \ (\alpha = 1,2,3,4) \qquad\qquad \frac{\partial f_\alpha}{\partial x_1} - f_{1\alpha} = 0 \ (\alpha = 1,2,3,4)$$

$$\frac{\partial f_{1\alpha}}{\partial x_2} - \frac{\partial f_{2\alpha}}{\partial x_1} = 0 \ (\alpha = 1,2,3,4)$$

(*) I don't give α the value 1 because the equations would not then be all independent. The same thing holds for $\alpha = 2$ in the forth line.

$$r_4 = 4 + 4 + 3 + 2 = 13, \quad r_3 = 4 + 4 + 3 = 11,$$
$$r_2 = 4 + 4 = 8, \quad r_1 = 4.$$

On the other hand, the number of unknown functions is

$$4 \text{ (for the } f_\alpha) + 10 = 14.$$
$$\mathscr{I}_4 = 14 - r_4 = 1, \quad \mathscr{I}_3 = 14 - 11 = 3,$$
$$\mathscr{I}_2 = 14 - 8 = 6, \quad \mathscr{I}_1 = 14 - 4 = 10,$$

instead of

$$\mathscr{I}_4 = 1, \quad \mathscr{I}_3 = 2, \quad \mathscr{I}_2 = 3, \quad \mathscr{I}_1 = 4.$$

La valeur de $\mathscr{I}_4$ n'a pas changé: c'est la seule qui ait une signification essentielle; les valeurs des autres $\mathscr{I}$ ne correspondent à aucune propriété intrinsèque du système, mais uniquement à la forme analytique particulière qu'on lui a donnée.

Si on avait eu $\mathscr{I}_4 = 0$, $\mathscr{I}_3 > 0$, on trouverait toujours $\mathscr{I}_4 = 0$ et la même valeur pour $\mathscr{I}_3$.

Vous remarquerez que conformément à la formule générale que je vous avais indiquée, en prolongeant mon système du 1er au second ordre, les nouvelles valeurs de $\mathscr{I}_1$, $\mathscr{I}_2$, $\mathscr{I}_3$, $\mathscr{I}_4$ sont

$$\mathscr{I}_1 + \mathscr{I}_2 + \mathscr{I}_3 + \mathscr{I}_4, \quad \mathscr{I}_2 + \mathscr{I}_3 + \mathscr{I}_4, \quad \mathscr{I}_3 + \mathscr{I}_4, \quad \mathscr{I}_4$$

$$4 + 3 + 2 + 1 = 10, \quad 3 + 2 + 1 = 6, \quad 2 + 1 = 3, \quad 1 \; ,$$

J'ai fait aux physiciens trois conférences qui ont semblé les intéresser [1]. Au sujet des conditions de compatibilité, M. Hadamard a fait deux remarques extrêmement importantes et qui vous intéresseront peut-être.

Dans ma théorie des systèmes en involution il est essentiel de supposer qu'on a des équations *analytiques*, que les solutions à une dimension qui déterminent une solution à deux dimensions sont *analytiques*, etc. Autrement dit le théorème de Cauchy-Kowalewsky sur l'existence d'une solution du système

$$\frac{\partial f_\alpha}{\partial z} = a_{\alpha\beta} \frac{\partial f_\beta}{\partial x} + b_{\alpha\beta} \frac{\partial f_\beta}{\partial y} + c_\alpha$$

correspondant à des valeurs initiales données $f_\alpha = \phi_\alpha(x,y)$ pour $z = 0$ suppose essentiellement que les données $\phi_\alpha(x,y)$ soient *analytiques*: dans ce cas on est assuré d'avoir pour les f_α des fonctions *analytiques* de x, y, z développables en séries convergentes. Mais ce théorème peut être en défaut si les données *ne sont pas* analytiques; il se peut que le système n'admette *aucune* solution correspondant à ces données. Un exemple simple est fourni par l'équation

$$\frac{\partial^2 f}{\partial z^2} + \frac{\partial^2 f}{\partial x^2} + \frac{\partial^2 f}{\partial y^2} = 0;$$

---

1. Voir note Lettre XXI.

186

The value of $\mathscr{I}_4$ has not changed: it is the only one that has an essential significance; the values of the other $\mathscr{I}$'s do not correspond to any intrinsic property of the system, but only to its special analytical form.

If $\mathscr{I}_4$ had been zero and $\mathscr{I}_3 > 0$, one would always find $\mathscr{I}_4 = 0$ and the same value for $\mathscr{I}_3$. You will note that, according to the general formula I indicated to you, by extending my system from the first to the second order, the new values of $\mathscr{I}_1$, $\mathscr{I}_2$, $\mathscr{I}_3$, $\mathscr{I}_4$ are

$$\mathscr{I}_1 + \mathscr{I}_2 + \mathscr{I}_3 + \mathscr{I}_4, \quad \mathscr{I}_2 + \mathscr{I}_3 + \mathscr{I}_4, \quad \mathscr{I}_2 + \mathscr{I}_3, \quad \mathscr{I}_4$$
$$4 + 3 + 2 + 1 = 10, \quad 3 + 2 + 1 = 6, \quad 2 + 1 = 3, \quad 1.$$

I have given three lectures to physicits that seem to have interested them [1]. Concerning the compatibility conditions, M. Hamadard has made two extremely important remarks that might perhaps interest you.

In my theory of systems in involution, it is essential to assume that the equations are *analytic*, that the one-dimensional solutions determining a two-dimensional solution are *analytic*, etc. In other words, the theorem of Cauchy-Kowalewsky on the existence of a solution for the system

$$\frac{\partial f_\alpha}{\partial z} = a_{\alpha\beta} \frac{\partial f_\beta}{\partial x} + b_{\alpha\beta} \frac{\partial f_\beta}{\partial y} + c_\alpha$$

corresponding to given initial values $f_\alpha = \phi_\alpha(x,y)$ for $z = 0$, essentially assumes that the data $\phi_\alpha(x,y)$ are *analytic*. In this case, one is sure to have *analytic* functions of x, y, z developable in convergent series for f_α. But this theorem may fail if the data *are not* analytic; it may happen that the system then allows for *no solution* corresponding to these data. A simple example is given by the equation

$$\frac{\partial^2 f}{\partial z^2} + \frac{\partial^2 f}{\partial x^2} + \frac{\partial^2 f}{\partial y^2} = 0;$$

187

si l'on se donne, pour $z = 0$,

$$f = \Phi(x,y), \ \frac{\partial f}{\partial z} = \Psi(x,y);$$

et si $\phi(x,y)$ n'est pas une fonction analytique de x, y, il est évident que l'équation ne peut admettre aucune solution correspondant à de telles données puisque toute fonction harmonique est analytique. Mais le théorème serait vrai pour l'équation

$$\frac{\partial^2 f}{\partial z^2} - \frac{\partial^2 f}{\partial x^2} - \frac{\partial^2 f}{\partial y^2} = 0.$$

En réalité la physique n'a jamais à considérer que des équations pour lesquelles la difficulté signalée par M. Hadamard ne se présente pas. Mais c'est tout de même une question importante qui est ainsi soulevée.

La seconde observation de M. Hadamard est aussi très intéressante. Considérons un système différentiel qui satisfasse au déterminisme, c'est-à-dire tel que la connaissance d'une solution soit déterminée par la valeur de la solution dans une section $x^4 = a$. Il peut arriver que si on se donne la solution dans cette section, la solution dans une section voisine soit pratiquement *inobservable* par le physicien: si on fait varier infiniment peu, *aussi peu qu'on veut,* les données dans la section $x^4 = a$, il peut arriver que dans une section $x^4 = a + \varepsilon$ aussi voisine qu'on veut de la première, il soit impossible de borner l'amplitude des variations que prendra la solution. M. Hadamard cite comme exemple le cas de l'équation

$$\frac{\partial^2 f}{\partial x^2} + \frac{\partial^2 f}{\partial t^2} = 0;$$

donnons-nous, pour $t = 0$,

$$f = \varepsilon \sin mx, \ \frac{\partial f}{\partial t} = 0,$$

ε étant une constante très petite. On aura ici

$$f = \varepsilon \sin mx \ ch \ mt.$$

On voit que, quelque petit que soit t, la fonction f observée à cet instant t peut prendre des valeurs aussi grandes qu'on veut (à cause de $ch \ mt$), bien que, dans la section $t = 0$, f reste partout très petit.

If one gives, at $z = 0$

$$f = \Phi(x,y), \frac{\partial f}{\partial z} = \Psi(x,y);$$

and if $\phi(x,y)$ is not analytic in x, y, it is obvious that the equation cannot allow for any solution corresponding to such data, since any harmonic function is analytic. But the theorem would be true for the equations

$$\frac{\partial^2 f}{\partial z^2} - \frac{\partial^2 f}{\partial x^2} - \frac{\partial^2 f}{\partial y^2} = 0.$$

Actually, physics need never consider equations for which the difficulty raised by M. Hadamard occurs. But he still has raised an important question.

M. Hadamard's second remark is also very interesting. Let us consider a deterministic differential system, i.e., a system such that the knowledge of a solution is determined by the value of the solution in a section $x^4 = a$. It may happen that the solution given in this section becomes in practice *unobservable* for the physicist in a near-by section; if, in the section $x^4 = a$, the data are varied infinitesimally, *as little as one pleases*, it may happen that, in a section $x^4 = a + \varepsilon$ as near as one pleases to the first one, it is impossible to limit the amplitude of the variations of the solution.

M. Hadamard quotes the example of the equation

$$\frac{\partial^2 f}{\partial x^2} + \frac{\partial^2 f}{\partial t^2} = 0;$$

Let us take, at $t = 0$,

$$f = \varepsilon \sin mx, \frac{\partial f}{\partial t} = 0,$$

ε being a very small constant. here we shall have

$$f = \varepsilon \sin mx \, ch \, mt.$$

Hence, however small t may be, the function f observed at that moment can take as large a value as one pleases (because of $ch \, mt$), although, in the section $t = 0$, f remains everywhere very small.

189

(Il y aurait donc dans certains cas un déterminisme *mathématique* qui ne serait pas à proprement parler un déterminisme *physique*.)

Le fait précédent ne se produirait pas si on considérait l'équation hyperbolique

$$\frac{\partial^2 f}{\partial x^2} - \frac{\partial^2 f}{\partial t^2} = 0.$$

Plus généralement l'équation

$$\frac{\partial^2 f}{\partial x^2} + \frac{\partial^2 f}{\partial y^2} + \frac{\partial^2 f}{\partial z^2} - \frac{1}{c^2} \frac{\partial^2 f}{\partial t^2} = 0$$

jouit de la propriété suivante. Si on prend une section d'Univers qui ne coupe pas le cône $dx^2 + dy^2 + dz^2 - c^2 dt^2 = 0$ (par ex. $t = a$), il y a, pour cette section d'Univers, un vrai déterminisme physique, mais il n'en est plus de même si la section d'Univers coupe ce cône (par ex. $z = c^{\text{te}}$).

Si l'on avait en Physique une équation de la forme

$$\frac{\partial^2 f}{\partial x^2} + \frac{\partial^2 f}{\partial y^2} - \frac{\partial^2 f}{\partial z^2} - \frac{\partial^2 f}{\partial t^2} = 0,$$

il n'y aurait déterminisme physique pour aucune section d'Univers. Heureusement que les équations de cette nature n'interviennent jamais en Physique.

Tout cela prouve que, dans l'état actuel de l'Analyse, on est obligé, dans la discussion de systèmes aussi compliqués que le vôtre, de se borner aux solutions analytiques. Du reste les caractéristiques étant les mêmes que pour l'équation de la propagation de la lumière, il ne doit pas y avoir de trop grandes craintes au sujet du déterminisme physique.

Je n'ai pas très bien compris ce que vous voulez dire au sujet du système $R_{ik} = 0$, etc. J'ai toujours considéré les R_{ik} comme se rapportant aux valeurs

$$g_{\mu\nu} = h_{s\mu} h_{s\nu}$$

et non pas

$$g_{\mu\nu} = \psi h_{s\mu} h_{s\nu}.$$

Du reste la fonction ψ ne peut s'introduire que si l'on prend

$$\frac{\partial \phi_\alpha}{\partial x^\beta} - \frac{\partial \phi_\beta}{\partial x^\alpha} = 0;$$

(Thus, in some cases, there seems to exist a *mathematical* determinism that is not, properly speaking, a *physical* determinism.)

The previous fact would not occur if one considered the hyperbolic equation

$$\frac{\partial^2 f}{\partial x^2} - \frac{\partial^2 f}{\partial t^2} = 0.$$

More generally, the equation

$$\frac{\partial^2 f}{\partial x^2} + \frac{\partial^2 f}{\partial y^2} + \frac{\partial^2 f}{\partial z^2} - \frac{1}{c^2}\frac{\partial^2 f}{\partial t^2} = 0$$

has the following property: If one takes a section of the Universe that does not intersect the cone $dx^2 + dy^2 + dz^2 - c^2 dt^2 = 0$ (e.g. $t = a$), there is, for this section of the Universe, a real physical determinism but this is no longer true if the section of the Universe cuts the cone (e.g. $z = const.$).

In Physics if one had an equation of the form

$$\frac{\partial^2 f}{\partial x^2} + \frac{\partial^2 f}{\partial y^2} - \frac{\partial^2 f}{\partial z^2} - \frac{\partial^2 f}{\partial t^2} = 0,$$

physical determinism would not exist for any section of the Universe. Happily, equations of that nature never occur in physics.

All this goes to prove that, in the present state of Analysis, and in the discussion of systems as complicated as yours, one is obliged to stick to analytic solutions. In any case, since the characteristics are the same for the equation of propagation of light, there are no fears about physical determinism.

I don't quite understand what you mean on the subject of the system $R_{ik} = 0$, etc. I have always considered the R_{ik} to refer to the quantities

$$g_{\mu\nu} = h_{s\mu}h_{s\nu}$$

and not to

$$g_{\mu\nu} = \psi h_{s\mu}h_{s\nu}.$$

Besides, the function ψ can only be introduced if one takes

$$\frac{\partial \phi_\alpha}{\partial x^\beta} - \frac{\partial \phi_\beta}{\partial x^\alpha} = 0;$$

on pourrait prendre plus généralement

$$\frac{\partial \phi_\alpha}{\partial x^\beta} - \frac{\partial \phi_\beta}{\partial x^\alpha} = a(\phi_\alpha S_\beta - \phi_\beta S_\alpha),$$

$$\frac{\partial S_\alpha}{\partial x^\beta} - \frac{\partial S_\beta}{\partial x^\alpha} = b(\phi_\alpha S_\beta - \phi_\beta S_\alpha)$$

avec deux constantes arbitraires a et b.

Avez-vous réfléchi à la question soulevée par Francis Perrin, à savoir la possibilité de compléter les seconds membres de vos équations de la manière suivante

$$\Lambda^\mu_{\alpha\beta;\mu} = 0,$$

$$G_{\alpha\beta} = k g_{\alpha\beta},$$

k étant une constante qu'on peut du reste supposer égale à 1 ou à -1? Cette modification revient à admettre l'existence d'une unité d'intervalle privilégiée. On ne trouve pas du reste avec cette modification de solution homogène fournissant un espace fini, mais F. Perrin a trouvé une solution régulière partout, non homogène, dans un espace sphérique avec temps infini [2].

Veuillez agréer, cher et illustre Maître, l'expression de mes sentiments tout dévoués.

E. Cartan

---

2. Voir note Lettre XXII.

More generally, one could take

$$\frac{\partial \phi_\alpha}{\partial x^\beta} - \frac{\partial \phi_\beta}{\partial x^\alpha} = a(\phi_\alpha S_\beta - \phi_\beta S_\alpha),$$

$$\frac{\partial S_\alpha}{\partial x^\beta} - \frac{\partial S_\beta}{\partial x^\alpha} = b(\phi_\alpha S_\beta - \phi_\beta S_\alpha)$$

with two arbitrary constants a and b.

Have you thought about the question raised by Francis Perrin, namely, the possibility of completing the left hand sides of your equations in the following way,

$$\Lambda^\mu_{\alpha\beta;\mu} = 0,$$

$$G_{\alpha\beta} = k g_{\alpha\beta},$$

k being a constant which in any case can be set equal to 1 or -1. This modification amounts to allowing the existence of a privileged unit of interval. Now with this modification, there is no homogeneous solution giving a finite space, but F. Perrin has found an inhomogeneous solution that is everywhere regular in a spherical space with indefinite time [2].

E. Cartan

XXXI

13-II-30

Verehrter Herr Cartan!

Wie gerne hätte ich Ihre drei Vorträge gehört! Aber ich bin glücklich darüber, dass ich so Interessantes von Ihnen brieflich erhalte. Jeder solche Brief ist eine wirkliche Freude für mich. Aber mit den $\mathscr{I}$ gebe ich mich immer noch nicht zufrieden. Ihr Argument kann nämlich gegen Sie selbst angewendet werden.

Denn Sie betrachten doch $\mathscr{I}_3$ bei einem vierdimensionalen Raum als Mass für die *généralité* (nicht $\mathscr{I}_4$). $\mathscr{I}_3$ ist aber für das System

$$\frac{\partial \phi_\alpha}{\partial x_\beta} - \frac{\partial \phi_\beta}{\partial x^\alpha} = 0$$

gleich 2; d.h. es können zwei der ϕ_α auf einem dreidimensionalen Schnitte frei gewählt werden, trotzdem alles durch *eine* willkürliche Funktion bestimmt ist. (Bei Ihrer Darstellung

$$\frac{\partial \phi_\alpha}{\partial x^\beta} = \phi_{\alpha\beta} \ (\phi_{\alpha\beta} = \phi_{\beta\alpha})$$

$$\phi_{\alpha\beta,\gamma} = \phi_{\alpha\gamma,\beta}$$

ist sogar $\mathscr{I}_3 = 3$).

Dies schadet aber nicht. Denn es wird diejenige Formulierung die richtige sein, in welchem die $\mathscr{I}_4, \mathscr{I}_3, \mathscr{I}_2, \mathscr{I}_1$ möglichst klein sind. Ich bin also immer noch der Ansicht, dass die *généralité* sachgemäss nur durch alle 4 Zahlen beschreiben werden soll. (Zörnen Sie mir nicht über meine Hartnäckigkeit; ich kann nicht anders). Ich bin sehr froh, Ihre Theorie der Involutions-Systeme nun ganz verstanden zu haben. Am Schönsten finde ich Ihren Beweis der Existenz der Identitäten. Ich freue mich auf Ihre ausführliche Publikation, die eine wichtige Frage endgültig klären wird.

Die erste der Hadamard-schen Bemerkungen hat mich sehr interessiert, aber nicht völlig überzeugt. Es scheint mir nämlich, dass man durch unstetige Verteilung von Massen *in der Fläche*

194

A XXXI

13.II.30

Dear M. Cartan

How delighted I would have been to have heard your three lectures! But I am lucky to have received so interesting a series of letters from you. Every such letter is truly a joy for me. But I am still not content with the $\mathscr{I}$'s. Your argument can, in fact, be turned against yourself.

For you consider $\mathscr{I}_3$ in a four dimensional space to be a measure of *généralité* (and not $\mathscr{I}_4$). But in the system

$$\frac{\partial \phi_\alpha}{\partial x_\beta} - \frac{\partial \phi_\beta}{\partial x^\alpha} = 0$$

$\mathscr{I}_3$ is equal to 2; two of the ϕ_α can be freely chosen on a three dimensional section, in spite of the fact that everything is fixed by *one* arbitrary function. (In your example,

$$\frac{\partial \phi_\alpha}{\partial x^\beta} = \phi_{\alpha\beta} \quad (\phi_{\alpha\beta} = \phi_{\beta\alpha})$$

$$\phi_{\alpha\beta,\gamma} = \phi_{\alpha\gamma,\beta}$$

$\mathscr{I}_3 = 3$).

But this doesn't matter. For that formulation will be right in which the $\mathscr{I}_4$, $\mathscr{I}_3$, $\mathscr{I}_2$, $\mathscr{I}_1$ are as small as possible. Thus I am still of the opinion that *généralité* can only be adequately described by all 4 numbers. (Do not be angry with me about my stubborness; I can do no other). I am very glad that I now fully understand your theory of systems in involution. The most beautiful part of it, I think, is your proof of the existence of identities. I am pleased with your detailed article which definitively clears up an important question.

The first of Hadamard's remarks interested me very much, but has not fully convinced me. In particular, it appears to me that, by means of a discontinuous distribution of masses *in the surface* $x_3 = const.$, one can produce functions $f\left(\dfrac{\partial^2 f}{\partial x^2} + \dfrac{\partial^2 f}{\partial y^2} + \dfrac{\partial^2 f}{\partial z^2} = 0\right)$

195

$x_3 = $ konst. Funktionen $f\left(\dfrac{\partial^2 f}{\partial x^2} + \dfrac{\partial^2 f}{\partial y^2} + \dfrac{\partial^2 f}{\partial z^2} = 0\right)$ erzeugen kann, welche in der Fläche $x_3 = $ konst. beliebige Unstetigkeiten besitzen (durch Doppel-Schichten, z.B. so

$$\begin{array}{c|c} & \\ \hline + & - \\ + & - \\ + & - \end{array} \quad x_3 = a).$$

Die zweite Bemerkung hat mich eigentlich nicht überrascht.

Ich wusste schon, dass in Ihrem Gleichungssystem die R_{ik} aus den $h_{s\mu}h_{sv}$ gebildet werden sollten und nicht aus den $\psi h_{s\mu}h_{sv}$. Ich wollte nur sagen, dass die aus den letzteren Grössen als $g_{\mu v}$ gebildeten R_{ik} mit meinem $G^{ik} + G^{ki}$ nahe verwandt zu sein scheinen.

Ist Ihnen deutlich geworden, warum mir Ihr System physikalisch nicht plausibel ist? Es ist, weil die g_{ik} durch $R_{ik} = 0$ schon deterministisch bestimmt werden, ohne dass die restlichen 6 Variabeln des Feldes eingehen. Diese erscheinen vielmehr nur a posteriori angehängt, ohne dass sie auf die g_{ik} zurückwirken. Ist dies klar?

F. Perrin's System mit dem kosmologischen Gliede könnte vielleicht noch einmal Bedeutung gewinnen. Einstweilen aber habe ich hierüber noch gar kein Urteil. Brennend ist verläufig die Frage nach der Existenz von singularitätsfreien Lösungen, welche die Elektronen und Protonen darstellen könnten. Denn ohne die Lösung dieses schwierigen Problems lässt sich über die Brauchbarkeit der Theorie nach meiner Meinung kein Urteil gewinnen. Einstweilen komme ich mir mit dieser Theorie vor wie ein ausgehungerter Affe, der nach langen Suchen eine ungeheure Kokosnuss gefunden hat, sie aber nicht öffnen kann; sodass er nicht einmal weiss, ob etwas darin ist.

Herzlich grüsst Sie

Ihr

A. Einstein

---

1. Voir note Lettre XXII.

which possess arbitrary discontinuities in the surface $x_3 = const.$ (e.g. by means of double layers

$$\begin{array}{c}
\underline{}\!\!\!\!\!\!-x_3 = a).\\
\begin{array}{c|c}
+ & - \\
+ & - \\
+ & -
\end{array}
\end{array}$$

The second remark did not really surprise me.

I knew, of course, that, in your system of equations, the R_{ik} are to be constructed out of the $h_{s\nu}h_{s\nu}$ and not out of $\psi h_{s\mu}h_{s\nu}$. I only wanted to say that the R_{ik} built out of the latter quantities, considered as $g_{\mu\nu}$, appear to be closely related to my $G^{ik} + G^{ki}$.

Has it become clear to you why your system is not physically plausible to me? It is because the g_{ik} are already deterministically fixed by $R_{ik} = 0$, without the remaining 6 field variables entering in. They appear to be added on in a rather *a posteriori* way, with no reaction on the g_{ik}. Is this clear?

F. Perrin's system, with the cosmological term, could perhaps, earn some support. But, for the present, I have no opinion on it. At this moment the burning question is that of the existence of singularity-free solutions which could represent electrons and protons. For without the solution of this difficult problem I feel no judgement can be passed on the usefulness of the theory. For the moment, this theory seems to me to be like a starved ape who, after a long search, has found an amazing coconut, but cannot open it; so he doesn't even know whether there is anything inside.

Kind regards.

Yours,

A. Einstein

XXXII

Le Chesnay (Seine et Oise)
27 avenue de Montespan,
le 17 février 1930 [1]

Cher et illustre Maître,

Je suis très fier que mes lettres puissent vous intéresser. Soyez sûr que de mon côté je regarde comme un privilège que vous vouliez bien me consacrer quelques-uns de vos instants, si précieux pour la science.

Je pense que, par une suite d'approximations successives, nous arriverons à nous entendre, même au sujet des $\mathscr{I}$. L'argument que vous me renvoyez contre moi-même m'a bien amusé, mais je refuse de le recevoir! En effet $\mathscr{I}_3$ mesure pour moi le degré de généralité *dans le cas où $\mathscr{I}_4$ est nul*, et c'est le cas de votre système et des systèmes analogues. Sinon c'est l'entier $\mathscr{I}_4$ qui indique le nombre de fonctions arbitraires de 4 variables qui mesure le degré de généralité de la solution: si $\mathscr{I}_4 > 0$, l'entier $\mathscr{I}_3$ n'a plus aucune signification essentielle. Si $\mathscr{I}_4 = 0$, c'est $\mathscr{I}_3$ que je prends comme mesure ($\mathscr{I}_3$ fonctions arbitraires de 3 variables). Si $\mathscr{I}_4 = \mathscr{I}_3 = 0$ c'est $\mathscr{I}_2$: la solution générale dépend alors de $\mathscr{I}_2$ fonctions arbitraires de 2 variables, et ainsi de suite. Si $\mathscr{I}_4 = \mathscr{I}_3 = \mathscr{I}_2 = \mathscr{I}_1 = 0$ la solution du système dépend d'un certain nombre $\mathscr{I}_0$ de *constantes* arbitraires, et ce nombre dans ce cas mesure aussi le degré d'indétermination de la solution.

Au reste je suis assez tenté de vous donner raison s'il s'agit de comparer entre eux deux systèmes de forme analytique semblable, comportant par exemple le même nombre d'équations aux dérivées partielles du même ordre au même nombre de fonctions inconnues, comme c'est le cas pour les deux systèmes possibles de 22 équations.

J'avais bien compris les raisons qui vous faisaient rejeter le système avec $R_{ik} = 0$; mais j'avais cru que dans une de vos dernières

---

1. La correspondance va s'interrompre ici pendant plus d'une année. On peut donner une première conclusion grâce à un billet d'Einstein à son ami Solovine. En date du 4 mars 1930 il écrit: « Ma théorie du champ fait de bons progrès. Cartan a

A XXXII

Le Chesnay (Seine et Oise)
27 avenue de Montespan
17 February 1930 [1]

Cher et illustre Maître,

I'm very proud my letters may be of some interest to you. Be sure that, for my part, I consider it to be a privilege that you are willing to spare me some of your time which is so precious for science.

I think that by successive approximations we shall arrive at an agreement, even about the $\mathscr{I}$'s. The argument that you turned against me amused me but I refuse to accept it! Indeed, $\mathscr{I}_3$ does measure, for me, the degree of generality *when $\mathscr{I}_4$ is zero*, and this is the case for your system and for similar systems. If this is not the case, it is the integer $\mathscr{I}_4$ that indicates the number of arbitrary functions of 4 variables measuring the degree of generality of the solution: if $\mathscr{I}_4 > 0$, the integer $\mathscr{I}_3$ no longer has an essential meaning. If $\mathscr{I}_4 = 0$, it is $\mathscr{I}_3$ that I take as a measure ($\mathscr{I}_3$ arbitrary functions of 3 variables). If $\mathscr{I}_4 = \mathscr{I}_3 = 0$, it is $\mathscr{I}_2$, the general solution then depending on $\mathscr{I}_2$ arbitrary functions of 2 variables, and so on. If $\mathscr{I}_4 = \mathscr{I}_3 = \mathscr{I}_2 = \mathscr{I}_1 = 0$, the solution of the system depends on a certain number $\mathscr{I}_0$ of arbitrary *constants*, and, in this case, this number also measures the degree of indeterminism of the solution.

For the rest, I am rather tempted to admit that you are right when it's a matter of comparing two systems of similar analytical form, involving the same number of partial differential equations, of the same order, and with the same number of unknown functions; as is the case for the two possible systems of 22 equations.

I quite understand you reasons for rejecting the system with $R_{ik} = 0$; but I thought that in one of your last letters you had partly

---

bien travaillé dans ce domaine. Moi-même je travaille avec un mathématicien (W. Mayer de Vienne), un splendide bonhomme qui occuperait depuis longtemps une chaire de professeur s'il n'était pas juif ». (*Lettres à Maurice Solovine*, loc. cit.)

lettres vous étiez revenu en partie sur votre opinion, à cause de la présence des équations $\frac{\partial S_\alpha}{\partial x^\beta} - \frac{\partial S_\beta}{\partial x^\alpha} = \dots$ qui font intervenir la métrique et le parallélisme. Mais évidemment cette dernière circonstance n'infirme en rien les arguments que vous m'aviez donnés et que vous vous êtes donné la peine de m'exposer à nouveau.

La première remarque d'Hadamard, relative à l'équation $\frac{\partial^2 f}{\partial x^2} + \frac{\partial^2 f}{\partial y^2} + \frac{\partial^2 f}{\partial z^2} = 0$, ne se rapporte pas à des données *discontinues* dans la section $z = 0$. D'une manière précise il n'existe *aucune* fonction f satisfaisant à l'équation considérée, et régulière dans l'intervalle $-\varepsilon < z < \varepsilon$, si l'on assujettit cette fonction à prendre pour $z = 0$ la valeur $f = \phi(x,y)$, la fonction ϕ étant *continue*, admettant même des dérivées des deux premiers ordres continues, mais *n'étant pas analytique*. Cela tient tout simplement à ce que toute fonction harmonique f régulière dans un domaine donné est nécessairement *analytique* dans ce domaine (analytique voulant dire développable en série de puissances); par suite la valeur que prend f sur la surface *analytique* $z = 0$ est nécessairement une fonction analytique de x, y et ne peut donc pas se réduire à $\phi(x,y)$.

Je comprends très bien que la solution trouvée par F. Perrin, de même que les solutions isolées que j'ai trouvées et qui sont dénuées de singularités, ne peuvent vous être d'aucun secours. (Tout le problème consiste à trouver une solution sans singularité assez générale pour qu'on puisse l'interpréter physiquement. Mais est-il bien sûr que de telles solutions existent et que la noix de coco contienne quelque chose à son intérieur?) On est devant un mur et les mathématiciens sont bien embarrassés pour y percer une ouverture. On ne peut guère fonder d'espoir que sur un miracle de divination; mais vous en avez déjà eu quelques-uns [2]!

Veuillez agréer, cher et illustre Maître, l'expression de mes sentiments tout à fait dévoués.

E. Cartan

---

2. Dans [12] après avoir exposé la difficulté relative aux variétés caractéristiques (cf. Note Lettre XII), E. Cartan conclut l'exposé général qu'il vient d'écrire sur le sujet en ces termes:

« On voit, par l'exposé qui précède, la variété des aspects sous lesquels peut être

reconsidered your opinion, because of the presence of the equations $\frac{\partial S_\alpha}{\partial x^\beta} - \frac{\partial S_\beta}{\partial x^\alpha} = \ldots$ that involve the metric and the parallelism. But of course, this last circumstance does not at all weaken the arguments that you took the trouble to explain to me again.

Hadamard's first remark, concerning the equation $\frac{\partial^2 f}{\partial x^2} + \frac{\partial^2 f}{\partial y^2} + \frac{\partial^2 f}{\partial z^2} = 0$, does not refer to *discontinous* data in the section $z=0$. To be precise, no function f, satisfying the equation and regular in the interval $-\varepsilon < z < \varepsilon$, exists if one constrains the function to take the value $f = \phi(x,y)$ at $z = 0$, the function ϕ being *continuous*, and even admitting continuous second derivatives, *but not being analytic*. This is simply due to the fact that any harmonic function f, regular in a given domain, is necessarily *analytic* in this domain (analytic meaning it can develop in a power series). As a result, the value taken by f on the analytic surface $z = 0$ is necessarily an analytic function of x, y and, so, cannot be reduced to $\phi(x,y)$.

I quite understand that the solutions found by F. Perrin, as well as the isolated singularity free solutions that I found, cannot be of any help to you. (The whole problem is to find a singularity-free solution general enough to be physically interpretable. But are we really sure that such solutions exist and that the coconut contains something inside?)

We find ourselve in front of a wall and we mathematicians are quite at a loss as to how to make a hole in it. One can only hope for some miracle of divination, but then you already have had several [2].

E. Cartan

---

envisagée la théorie unitaire du champ et aussi la difficulté des problèmes qu'elle soulève. Mais M. Einstein n'est pas de ceux à qui les difficultés font peur et, même si sa tentative n'aboutit pas, elle nous aura forcés à réfléchir sur les grandes questions qui sont à la base de la science. »

XXXIII

Caputh bei Potsdam,
den 13 Juni 1931

Verehrter Herr Cartan!

Ich habe mit grossem Interesse und Vergnügen Ihren schönen Aufsatz in der „ *Scientia* " [1] gelesen. In der Zwischenzeit habe ich zusammen mit Dr. Mayer viel über den Gegenstand gearbeitet und bin von den damaligen Feldgleichungen abgekommen. Nach jenen Feldgleichungen scheint nämlich keine Gravitationswirkung zu existieren, da statische Lösungen mit beliebig vielen Massenpunkten (Singularitäten) existieren, für welche nur h_{44} von 0 verschieden ist. Es gelang uns, systematisch alle Feldgleichungen vom ins Auge gefassten Typ abzuleiten, welche einer Identität genügen. Unter diesen ist eine besonders interessant, welche in der Arbeit mit I_{221} bezw. (11), (12) bezeichnet ist. Dieses System untersuchen wir nun genauer [24].

Es ist Ihnen vielleicht nicht entgangen, dass mein langjähriger Freund und früherer Arbeitsgenosse M. Grossmann in Zürich Ihre und meine Arbeiten kritisiert hat [28]. Herr Grossmann ist ein schwer leidender Mann, der in gesunden Tagen ein intelligenter Mathematiker war, heute aber augenscheinlich nicht mehr im Vollbesitz seiner geistigen Kräfte ist. Für den Fall, dass Sie das furchtbare Leiden meines Freundes irgendwie kennen, sage ich Ihnen, dass er an einer vorgeschrittenen Multiplisklerose leidet. Ich sage Ihnen all dies, um Ihnen nahezulegen, dass Sie ihm nicht öffentlich antworten. Ich habe in einem langen Briefe versucht, ihm seinen Irrtum in rücksichtsvoller Form darzulegen, habe aber wenig Hoffnung, von dem Bedauernswerten noch verstanden zu werden. Natürlich hängt die unhöfliche Form seiner Auseinandersetzung ebenfalls mit seiner Krankheit zusammen.

Herzlich grüsst Sie

Ihr

A. Einstein

---

1. Voir [11] et le post-scriptum de la lettre suivante.

A XXXIII

Caputh bei Potsdam
13 June 1931

Dear M. Cartan,

I have read with great interest and pleasure your beautiful article in *Scientia* [1]. In the meantime, I have been working a great deal with Dr. Mayer on the subject and I have abandoned those field equations. The reason is that it appears that, according to those field equations, there are no gravitational effects, since static solutions with arbitrarily many point masses (singularities) exist for which only h_{44} is non-zero. We have succeeded in systematically deriving all field equations of the required type which satisfy an identity. Among these there is one of especial interest, referred to in the work as I_{221} or (11), (12). We are now looking further into this system [24].

It has perhaps not escaped your notice that, in Zurich, my old friend and former coworker, M. Grossmann, has criticized your work and mine [28]. Herr Grossmann is a very ill man who, in healthier times, was a bright mathematician; but, today, evidently, is no longer in full possession of his intellectual faculties. In case you have not heard of the terrible illness of my friend I will tell you that he is suffering from an advanced case of multiple sclerosis. I tell you all this to urge you not to answer him publicly. In a long letter I have tried to point out his error to him in a courteous way, but I have little hope of making that unfortunate man understand. Naturally, the rude manner of his opposition is also a consequence of his illness.

Kind regards.

Yours,

A. Einstein

XXXIV

Le Chesnay (S. et O.)
27 avenue de Montespan,
le 24 juin 1931

Cher et illustre Maître,

M. Grossmann m'a en effet envoyé son mémoire. Je ne lui ai pas encore répondu — autrement qu'en lui envoyant mon article de la « *Revue de Métaphysique et de Morale* » sur la théorie unitaire [12]. Sans connaître l'état de santé de M. Grossmann, j'ai pensé qu'il n'y avait lieu de se livrer à une controverse publique, car je ne vois guère l'utilité de convaincre les lecteurs qui ne seraient pas déjà convaincus par la lecture du mémoire de M. Grossmann. J'avais même l'intention de vous écrire pour vous demander ce que vous pensiez de la chose lorsque votre lettre m'a prévenu. Mais tout cela est bien triste pour votre ami et pour vous-même. Je me demande seulement comment la rédaction du *Vierteljahrsschrift* a publié ce mémoire.

Je vous remercie de m'avoir envoyé un tirage à part de votre note des *Sitzungsberichte* [24]. Je m'étais moi-même l'année dernière livré [1] au travail de rechercher tous les systèmes en involution possibles conformes au déterminisme et, en dehors des deux systèmes de 22 équations, je n'avais vu comme possibles que des systèmes de 16 équations avec 4 identités. Mais j'avais reculé devant la longueur des calculs nécessaires pour trouver la forme des termes quadratiques. Si M. Mayer et vous arrivez à les simplifier, comme il semble bien que vous le faites, ce sera remarquable et je vous souhaite d'arriver ainsi à une théorie satisfaisante.

J'avais néanmoins trouvé — et je crois bien vous l'avoir indiqué — une catégorie très étendue de solutions possibles à 15 équations. Il faudrait supposer que les équations (1) de votre note se réduisent à 15 au lieu de 16:

$$a_1 + a_2 + 4a_3 = p + q, \quad R_\alpha^\alpha = 0,$$

---

1. Voir Note I, Lettre XXI, et [13].

204

A XXXIV

Le Chesnay (S. et O.)
27 avenue de Montespan,
24 June 1931

Cher et illustre Maître,

M. Grossman has indeed sent me his article. I have not yet replied — except by sending him my paper from *Revue de Métaphysique et de Morale* on the unified theory [12]. Although I did not know M. Grossman's state of health, I did think there was no reason to engage in a public debate, for I see no purpose in convincing readers who would not already have been convinced by reading M. Grossman's memoir. I even intended writing you to ask your opinion about it when your letter arrived. But all this is very sad for both your friend and yourself. I am only surprised that the editors of the *Vierteljahrschrift* published this article.

I thank you for having sent me the reprint of your article in the *Sitzungsberichte* [24]. Last year, I myself had set to work to find all the possible deterministic systems in involution [1] and, except for the two systems of 22 equations, there seemed to be only systems of 16 equations with 4 identities. But I recoiled from the lengthy calculations needed to find the form of the quadratic terms. If M. Mayer and you do manage to simplify them, as it seem you are doing, it will be remarkable, and I hope you may thus arrive at a satisfactory theory.

Nevertheless I did find — and I think I told you — a very large class of possible solutions with 15 equations. One would have to assume that equations (1) of your note reduce to 15 instead of 16:

$$a_1 + a_2 + 4a_3 = p + q; \quad R_\alpha^\alpha = 0,$$

and that the skew-symmetric part of the system $G^{\alpha\beta} - G^{\beta\alpha} = 0$ reduces to

$$S_{\alpha,\beta} - S_{\beta,\alpha} = 0.$$

et que la partie antisymétrique du système $G^{\alpha\beta} - G^{\beta\alpha} = 0$ se réduise à

$$S_{\alpha,\beta} - S_{\beta,\alpha} = 0.$$

Dans ce cas très général *où les $R^{\mu\alpha}$ sont quelconques, avec la seule restriction $R^{\alpha}_{\alpha} = 0$*, on a un système en involution conforme au déterminisme. Il comporte 4 identités du reste évidentes

$$(S_{\alpha,\beta} - S_{\beta,\alpha})_{,\gamma} + (S_{\beta,\gamma} - S_{\gamma,\beta})_{,\alpha} + (S_{\gamma,\alpha} - S_{\alpha,\gamma})_{,\beta} \equiv 0.$$

Le prochain fascicule du *Bulletin de la Société Mathématique de France* [13] contiendra un article où j'expose ma théorie des systèmes en involution sous la forme que je vous avais indiquée l'année dernière. Je me ferai un plaisir de vous en envoyer un tirage à part.

Veuillez agréer, cher et illustre Maître l'expression de mes sentiments cordialement dévoués.

E. Cartan

P.S. Vous me parlez dans votre lettre de mon article de « *Scientia* »; c'est sans doute de celui que j'ai publié sur la théorie unitaire dans *la Revue de Métaphysique et de Morale* [12]. Je viens également de publier dans *Scientia* un article: Géométrie euclidienne et géométrie riemannienne [11], mais qui n'a aucun rapport avec la physique; je vous l'adresse par le même courrier.

In this very general case, *where the* $R^{\mu\alpha}$ *are arbitrary with* $R^{\alpha}_{\alpha} = 0$ *as the sole restriction*, one has a deterministic system in involution. It allows for 4 obvious identities

$$(S_{\alpha,\beta} - S_{\beta,\alpha})_{,\gamma} + (S_{\beta,\gamma} - S_{\gamma,\beta})_{,\alpha} + (S_{\gamma,\alpha} - S_{\alpha,\gamma})_{,\beta} \equiv 0.$$

The next issue of the " *Bulletin de la Société Mathématique de France* " [13] will contain an article in which I outline my theory of systems in involution in the form I told you about last year. It will be a pleasure to send you a reprint.

E. Cartan

P.S. In you letter you mention my article in *Scientia*. This is, no doubt, the one I published on the unified theory in *Revue de Métaphysique et de Morale* [12]. I have also just published an article in *Scientia*: " *Géométrie euclidienne et géométrie riemannienne* " [11] but it has nothing to do with physics; I am sending it to you in the same post.

XXXV

An Bord. S. Franzisko [1]
den 21.III.32
Hamburg-Amerika Linie

Verehrter Herr Cartan !

Ich habe mit grossem Genuss Ihre Arbeit über Involutions-Systeme gelesen [13]. Dies scheint mir ein wirklich wichtiger Beitrag zur Theorie der partiellen Differenzial-Gleichungen zu sein.

Nun wäre es schön, wenn Sie den Satz ebenso sorgfältig behandeln wollten, dass ein Nicht-Involutions-System durch Differentiation und Einführung zusätzlicher abhängiger Variabeln stets ein Involutionssystem oder zu einem in sich widerspruchsvollen System führt. So scheint z.B. vom formalen Standpunkt aus das System

$$\Lambda^{\alpha}_{\mu\nu;\nu} = 0$$

ein ganz natürliches System von nicht involutorischen Charakter zu sein. Gibt es wirklich nur ganz triviale Lösungen so eines Systems? Es ist eigentlich schwer, dies zu glauben.

Übrigens bin ich jetzt von der Methode des Fern-Parallelismus gänzlich abgekommen. Es scheint, dass diese Struktur mit der wahren Beschaffenheit des Raumes nichts zu thun hat. Ich habe mit Dr. Mayer seit einem Jahre eine andere Theorie verfolgt, nämlich die der Fünfer-Vektoren und der Fünfer-Metrik im vierdimensionalen metrischen Kontinuum [2]. Diese Theorie liefert nicht nur ganz ungezwungen die Maxwell'schen Gleichungen sondern auch — wie ich in der letzten Zeit fand — eine Erweiterung der letzeren, welche kontinuierlich verteilte elektrische Massen zulässt. Ich habe Hoffnung, dass diese Theorie wirklich der Struktur des physikalischen Raumes näher kommt, ohne dass man, in Grundgesetzen eine nur statistische Interpretation gibt. Mit diesem Dogma der neuen Physiker-Generation

---

1. Einstein s'est embarqué à New York à son retour d'un second séjour à Pasadena.

21 MAY 1932

A XXXV

On Board the San Francisco [1]
21.III.32
Hamburg-America Line

Dear M. Cartan,

I have read with great enjoyment your work on systems in involution [13]. This seems to me to be a truly important contribution to the theory of partial differential equations.

It would be nice now if you were willing to deal equally carefully with the theorem that a system not in involution always leads, by differentiation and the introduction of supplementary dependent variables, to either a system in involution or a self-contradictory system. Thus, for example, it appears, from a formal point of view, that the system

$$\Lambda^{\alpha}_{\mu\underline{\nu};\nu} = 0$$

is a quite natural system of non-involutive nature. Are there really only completely trivial solutions of such a system? This is very hard to believe.

In any case, I have now completely given up the method of distant parallelism. It seems that this structure has nothing to do with the true character of space. For some years, together with Dr. Mayer, I have pursued another theory, that of the 5-vector and 5-metric in a four dimensional metric continuum [2]. This theory not only yields Maxwell's equations in a natural way but also — as I have recently discovered — an extension of them which admits continuously distributed charged masses. I have hopes that this theory really comes closer to the structure of physical space without its basic laws having to be given a merely statistical interpretation. I absolutely cannot reconcile myself to this dogma of the new generation of physicists, no

2. Voir Note lettre suivante.

kann ich mich nämlich absolut nicht befreunden, so verführerisch auch die Argumente sind, die man hiefür geltend macht. Sobald die neue Theorie ausgearbeitet ist, will ich Sie Ihrem Urteil unterbreiten.

Der Hauptgrund der Unbrauchbarkeit der Konstruktion des Fern-Parallelismus scheint darin zu liegen, dass man den in dieser Theorie existierenden „geraden Linien" absolut keine physikalische Bedeutung beilegen kann, während die physikalische bedeutungsvollen (makroskopischen) Bewegungsgleichungen nicht zu erhalten sind [3]. Mit andern Worten: die h_{sv} liefern keine brauchbare Darstellung des elektromagnetischen Feldes.

Herzlich grüsst Sie Ihr

A. Einstein

P.S. Noch eine Bemerkung zu Ihrer Arbeit. Es ist mir nicht recht klar, was Sie unter „ *degré d'arbitraire* " verstehen. Es scheint mir, dass man den Sachverhalt nicht einfacher ausdrücken kann als so:
Es bleiben unbestimmt bezw. frei wählbar:

$n - r + r_3$ Funktionen von 4 Variabeln,

$n - r_3 + r_2$ Funktionen von 3 Variabeln,

$n - r_2 + r_1$ Funktionen von 2 Variabeln,

$n - r_1$ Funktionen von 1 Variabel.

---

3. C'est effectivement l'impossibilité de déduire les équations du mouvement des équations de champ qui conduit Einstein à abandonner sa tentative.

matter how enticing the arguments are that are made for it. As soon as the new theory is worked out, I want to submit it to your judgement.

The main reason for the uselessness of the distant parallelism construction lies, I feel, in that one can attribute absolutely no physical meaning to the "straight lines" of the theory, while the physically meaningful (macroscopic) equations of motion cannot be obtained from it [3]. In other words, the h_{sv} give rise to no useful representation of the electromagnetic field.

Kind regards.

Yours,

A. Einstein

P.S. Another comment on your work. It is not entirely clear what you mean by "*degré d'arbitraire*". It seems to me that the facts of the matter cannot be more simply stated than as follows.

There remain unfixed, i.e. freely choosable

$$n - r + r_3 \quad \text{functions of 4 variables}$$
$$n - r_3 + r_2 \quad \text{functions of 3 variables}$$
$$n - r_2 + r_1 \quad \text{functions of 2 variables}$$
$$n - r_1 \quad \text{functions of 1 variable.}$$

XXXVI

Albert Einstein

Berlin W. den 26 April 1932
Haberlandstr. 5

Sehr geherter Herr Cartan !

Es hat mich tief berührt, dass Sie und Ihre Frau so schweres, tiefes Leid durchgemacht haben [1]. Wie schwer zu ertragen muss das Leben erst für solche sein, deren Leben sich im Persönlichen erschöpft und die keiner grossen, unpersönlichen Sache bis zum Selbstvergessen dienen können. Wie oft habe auch ich mich in die Befreiung durch objektiv gerichtete Beschäftigung gerettet!

Mein Mitarbeiter, der treffliche Herr Dr. Mayer lässt Ihnen herzlich für Ihre Arbeit danken, die er mit mir zusammen mit grossem Genuss gelesen hat. Er sendet Ihnen unsere beiden Arbeiten über Elektrizität, die zweite in der Korrektur [2]. Es sieht fast so aus, wie wenn wirklich etwas Gutes daran wäre... aber es gibt nur einen Weg der Wahrheit und unzählige des Irrens.

In dem „ *degré d'arbitraire* " kann ich immer noch nicht bestimmen. Um Ihnen mein Bedenken recht plastisch zu zeigen, nehme ich zwei Involutionssysteme vor, die sich nur durch das Weglassen einer Gleichung unterscheiden.

I. System $\frac{\partial u_3}{\partial x_2} - \frac{\partial u_2}{\partial x_3} = 0; \frac{\partial u_1}{\partial x_3} - \frac{\partial u_3}{\partial x_1} = 0; \frac{\partial u_2}{\partial x_1} - \frac{\partial u_1}{\partial x_2} = 0.$

II. System $\frac{\partial u_3}{\partial x_2} - \frac{\partial u_2}{\partial x_3} = 0; \frac{\partial u_1}{\partial x_3} - \frac{\partial u_3}{\partial x_1} = 0.$

---

1. Le 26 mars 1932 Elie Cartan perdait son fils Jean. C'était un compositeur de grand talent, malheureusement atteint par la tuberculose; il est mort en sanatorium à l'âge de 25 ans.

Lors des cérémonies organisées à l'occasion du jubilé scientifique d'Elie Cartan, le Comité d'organisation a fait exécuter une œuvre de Jean Cartan, « Hommage à Dante »; l'orchestre de la Société des Concerts de Paris était placé sous la direction de Charles Münch.

2. Il s'agit de [25] et [26].

Seul le premier article a été traduit en français par M. Solovine: *A. Einstein: Théorie de la relativité*, Hermann, Paris, 1933, pp. 73-98.

A XXXVI

Albert Einstein

Berlin W. 26 April 1932
Haberlandstr. 5

My dear M. Cartan,

I am deeply moved that you and your wife have experienced such a heavy and deep sorrow [1]. How difficult to endure must life be, at first, for someone whose life is led completely on the personal level and who can make use of no great, impersonal concern to the point of self-forgetfulness. How often have I too escaped into release through objective and ordered work.

My coworker, the excellent Dr. Mayer, thanks you very much for your article which he, and I, have read with great pleasure. He is sending you our two articles on electricity, the second one in proof [2]. It almost seems as if there might be something good in it ... but there is only one true path and so many false ones.

I still cannot agree with you on the subject of the " *degré d'arbitraire* ". In order to show you my ideas in a graphic way, I will take two systems in involution, which differ only by the absence of one equation.

System I $\dfrac{\partial u_3}{\partial x_2} - \dfrac{\partial u_2}{\partial x_3} = 0; \ \dfrac{\partial u_1}{\partial x_3} - \dfrac{\partial u_3}{\partial x_1} = 0; \ \dfrac{\partial u_2}{\partial x_1} - \dfrac{\partial u_1}{\partial x_2} = 0.$

System II $\dfrac{\partial u_3}{\partial x_2} - \dfrac{\partial u_2}{\partial x_3} = 0; \ \dfrac{\partial u_1}{\partial x_3} - \dfrac{\partial u_3}{\partial x_1} = 0.$

La lecture de cette théorie d'Einstein et Mayer a été l'objet d'un mémoire d'E. Cartan, curieusement resté à l'état de manuscrit et publié en 1955 par les éditeurs des *Œuvres Complètes d'E. Cartan* [15].

Dans la seconde [26] ,en dernière page (p. 137), après avoir présenté une discussion de la compatibilité des équations de champ, les auteurs soulignent l'effort d'E. Cartan dans les termes suivants: « Remarquons que M. Cartan, dans une recherche très générale et tout à fait éclairante, a analysé en profondeur cette propriété dite de compatibilité dans le présent travail ainsi que dans des travaux antérieurs », et ils donnent en note la référence à [13].

Beide Systeme sind offenbar Involutions-Systeme, das zweite, weil es überhaupt (bei allgemeiner Wahl der Koordinaten) keine durch Elimination von den $\frac{\partial u}{\partial x_3}$ gewinnbare Gleichung besitzt. Die Systeme sind durch die Grössen gekennzeichnet.

	n	r	r_2	r_1
I	3	3	1	0
II	3	2	0	0

Für den degré d'arbitraire kommt also in Betracht

	$n - r + r_2$	$n - r_2 + r_1$	$n - r_1$
I	1	2	3
II	1	3	3

Die beiden Fälle sind doch dem Sinne nach nicht äquivalent inbezug auf den degré d'arbitraire. Aber man kann den Unterschied nicht wohl ausdrücken, wenn man die Beschaffenheit durch eine einzige Zahl charakterisiert.

Das ist nur eine unbedeutende Einzelheit. Aber ich habe doch keine Ruhe, solange wir nicht einig sind; ich denke immer, ich habe Sie irgendwie missverstanden.

In Herzlichkeit und Sympathie grüsst Sie

Ihr

A. Einstein

Both systems are clearly systems in involution, the second because it contains no equation obtainable by elimination of the $\frac{\partial u}{\partial x_3}$ (for a general choice of coordinates). The systems may be characterized by the values

	n	r	r_2	r_1
I	3	3	1	0
II	3	2	0	0

Then, for the *degré d'arbitraire*, we have

	$n - r + r_2$	$n - r_2 + r_1$	$n - r_1$
I	1	2	3
II	1	3	3

The two cases are, therefore, not equivalent with respect to the *degré d'arbitraire*. But the difference cannot be well expressed if one tries to characterize properties by a single number.

This is only an unimportant peculiarity. However, I shall have no rest so long as we are not of the same mind; I always fear I have misunderstood you somehow. With sincere and sympathetic regards.

Yours,

A. Einstein

XXXVII

Le Chesnay, le 29 avril 1932

Cher et illustre Maître,

Je vous remercie bien vivement de votre sympathie dans notre grand malheur. Comme vous me le dites, ce n'est pas à soi-même qu'on doit penser.

Je n'ai pas encore reçu les tirages à part que vous m'annoncez, mais je ne veux pas attendre plus longtemps avant de répondre à votre observation sur le « degré d'arbitraire » de la solution générale d'un système différentiel.

Je remarque d'abord, pour dissiper toute équivoque, que la notion du degré d'indétermination de la solution d'un système comporte une grande part de convention. S'il s'agit d'équations différentielles ordinaires, par exemple d'une équation différentielle du n^e ordre à une fonction inconnue y d'une variable indépendante x, cela un sens très net de dire que la solution générale dépend de n constantes arbitraires. Mais prenons par exemple l'équation

$$\frac{\partial^2 z}{\partial y^2} - \frac{\partial z}{\partial x} = 0. \tag{1}$$

On peut dire que la solution générale (supposée *analytique*) dépend de deux fonctions arbitraires de x, à savoir les fonctions $f(x)$ et $\phi(x)$ auxquelles se réduisent pour $y = 0$ les fonctions z et $\frac{\partial z}{\partial y}$; ces fonctions étant données on a en effet

$$z = f(x) + y\phi(x) + \frac{y^2}{2} f'(x) + \frac{y^3}{3!} \phi'(x) + \dots \tag{2}$$

Si $f(x)$ et $\phi(x)$ sont données par leurs développements en séries, convergentes par exemple pour $|x| < R$, on obtient pour z un développement convergent au voisinage de $x = 0$, $y = 0$.

Mais on peut dire aussi que la solution générale de (1) ne dépend que d'*une* fonction arbitraire d'une variable, à savoir la fonc-

216

A XXXVII

Le Chesnay, 29 April 1932

Cher et illustre Maître,

My warmest thank for your sympathy in our great misfortune. As you say, it is not of oneself that one must think.

I haven't yet received the reprints you mentioned but I don't want to wait any longer before replying to your remark on the *degré d'arbitraire* of the general solution of a differential system.

In order to clear up any misunderstanding, I first remark that the concept of degree of indetermination of the solution of a system is, to a great extent, conventional. If it is a question of ordinary differential equations, for instance a differential equation of order n with one unknown function y of one independent variable x, then there is a very clear sense in which one can say that the general solution depends on n arbitrary constants. But let us consider the example of the equation

$$\frac{\partial^2 z}{\partial y^2} - \frac{\partial z}{\partial x} = 0. \tag{1}$$

One can say that the general solution (assumed to be *analytic*) depends on two arbitrary functions of x, namely, the functions $f(x)$ and $\phi(x)$, which are the restrictions of z and $\frac{\partial z}{\partial y}$ to $y = 0$. These functions being given, one has

$$z = f(x) + y\phi(x) + \frac{y^2}{2}f'(x) + \frac{y^3}{3!}\phi'(x) + \ldots \tag{2}$$

If $f(x)$ and $\phi(x)$ are given by their series expansions, convergent let us say for $|x| < R$, one obtains a series for z that converges in a neighbourhood of $x = 0$, $y = 0$.

But one can also say that the general solution of (1) depends on only *one* arbitrary function of one variable, namely, the function $\psi(y)$

217

tion $\psi(y)$ à laquelle z se réduit pour $x = 0$, car cette fonction étant connue, on a (toujours dans le cas d'une solution analytique)

$$z = \psi(y) + x\psi''(y) + \frac{x^2}{2!}\psi^{IV}(y) + \dots \qquad (3)$$

Encore faut-il remarquer que la fonction $\psi(y)$ n'est pas absolument arbitraire, parce que si on la prend sans précaution, le développement (3) de *z ne sera pas convergent*. En réalité donc la solution générale de (1) dépend d'*une seule fonction $\psi(y)$ assujettie à certaines conditions*. Ce résultat est, en apparence, complètement contradictoire avec le premier résultat (z dépend de *2* fonctions arbitraires $f(x)$, $\phi(x)$).

On peut encore ajouter que la donnée de deux fonctions arbitraires (analytiques) $f(x)$ et $\phi(x)$ est équivalente à la donnée d'*une seule* fonction arbitraire F(x). Supposons en effet $f(x)$ et $\phi(x)$ régulières au voisinage de $x = 0$ et posons

$$F(x) = f(x^2) + x\phi(x^2);$$

la connaissance de F(x) *entraîne celle de* $f(x)$ *et de* $\phi(x)$. Posons en effet

$$f(x) = a_0 + a_1 x + a_2 x^2 + \dots,$$
$$\phi(x) = b_0 + b_1 x + b_2 x^2 + \dots,$$

on a

$$F(x) = a_0 + b_0 x + a_1 x^2 + b_1 x^3 + a_2 x^4 + b_2 x^5 + \dots$$

d'où la démonstration de mon affirmation. On pourrait donc dire, au sens précédent, que la donnée de *deux* fonctions analytiques de x n'est pas plus générale que celle d'*une seule* fonction analytique de x.

Tout cela montre bien que le degré d'indétermination ou d'arbitraire des données qui déterminent la solution générale d'un système d'équations aux dérivées partielles comporte une grande part de convention. La convention que j'ai faite dans mon mémoire est légitime, en ce sens que, quel que soit le choix *non singulier* des variables, elle conduit toujours au même résultat. Cela est vrai, soit qu'on se borne, comme je l'ai fait, au premier nombre non nul de la suite

$$n - r + r_2, \; n - r_2 + r_1, \; n - r_1$$

soit qu'on prenne tous les nombres de cette suite.

Nous arrivons maintenant à la question en litige. Il est clair qu'en prenant tous les nombres de la suite, cela permet, au moins en

to which z reduces for $x = 0$, because if this function is known one has (again in the case of an analytic solution)

$$z = \psi(y) + x\psi''(y) + \frac{x^2}{2!}\psi^{IV}(y) + \dots \tag{3}$$

And one should notice also that the function $\psi(y)$ is not completely arbitrary, because of one chooses it without sufficient care, the expansion (3) *will not converge*. In reality, the general solution of (1) depends on *a single function $\psi(y)$ subject to certain conditions*. This result is, apparently, in complete contradiction with the first result (z depends on 2 arbitrary functions $f(x)$, $\phi(x)$).

One should add here that giving two arbitrary (analytic) functions $f(x)$ and $\phi(x)$ is equivalent to giving *a single* arbitrary function $F(x)$. Let us assume that $f(x)$ and $\phi(x)$ are regular in a neighbourhood of $x = 0$, and let us write

$$F(x) = f(x^2) + x\phi(x^2).$$

Then a knowledge of $F(x)$ *entails that of $f(x)$ and $\phi(x)$.* If we write

$$f(x) = a_0 + a_1 x + a_2 x^2 + \dots,$$
$$\phi(x) = b_0 + b_1 x + b_2 x^2 + \dots,$$

so one has

$$F(x) = a_0 + b_0 x + a_1 x^2 + b_1 x^3 + a_2 x^4 + b_2 x^5 + \dots$$

and this proves my statement. In this sense, one could say that giving *two* analytic functions of x is not more general than giving *a single* analytic function of x.

All this clearly shows that the degree of indetermination or arbitrariness of the data that determine the general solution of a system of partial differential equations contains a great deal of convention. The convention made in my article is legitimate, in the sense that, whatever the *non-singular* choice of the variables, one is always led to the same results. This is true whether one restricts oneself, as I have done, to the first non-zero number of the sequence

$$n - r + r_2, \; n - r_2 + r_1, \; n - r_1$$

or if one considers all the numbers of this sequence. We now come to the contentious question. It is clear that taking all the numbers of the

apparence, de distinguer les uns des autres des systèmes non équivalents entre eux. Les deux systèmes (I) et (II) de votre lettre, en un certain sens, n'ont pas le même degré de généralité, puisque la solution générale du premier est

$$u_1 = \frac{\partial F}{\partial x_1}, \; u_2 = \frac{\partial F}{\partial x_2}, \; u_3 = \frac{\partial F}{\partial x_3}$$

celle du second est

$$u_1 = \frac{\partial F}{\partial x_1} + \frac{\partial \phi}{\partial x_1}, \; u_2 = \frac{\partial F}{\partial x_2} - \frac{\partial \phi}{\partial x_2}, \; u_3 = \frac{\partial F}{\partial x_3},$$

F étant une fonction arbitraire de x_1, x_2, x_3, et ϕ une fonction arbitraire de x_1, x_2. À s'en tenir aux formules précédentes, la solution générale de (I) dépend d'une fonction arbitraire de 3 variables, celle de (II) d'une fonction arbitraire de 3 variables *et* d'une fonction arbitraire de 2 variables (l'un et l'autre énoncé étant du reste en désaccord avec les valeurs trouvées pour $n - r_2 + r_1$ et $n - r_1$!).

J'admets donc très volontiers avec vous que le premier nombre non nul de la suite $n - r + r_2$, $n - r_2 + r_1$, $n - r_1$ est tout à fait insuffisant pour donner une idée précise de cet élément, du reste si fuyant, qu'est le degré de généralité de la solution d'un système. Pourquoi est-ce que je me suis borné à ce seul nombre? C'est parce que les autres ne se conservent pas toujours quand on remplace le système par un autre équivalent. J'appelle systèmes équivalents deux systèmes au même nombre de variables indépendantes tels qu'il existe une correspondance biunivoque entre les solutions de ces deux systèmes. Je prends par exemple votre Système I

$$\frac{\partial u_3}{\partial x_2} - \frac{\partial u_2}{\partial x_3} = 0, \; \frac{\partial u_1}{\partial x_3} - \frac{\partial u_3}{\partial x_1} = 0, \; \frac{\partial u_2}{\partial x_1} - \frac{\partial u_1}{\partial x_2} = 0,$$

à 3 fonctions u_1, u_2, u_3 de x_1, x_2, x_3. Je considère maintenant le système suivant à 9 fonctions inconnues u_i et $u_{ij} = u_{ji}$ ($i, j = 1, 2, 3$):

$$\begin{cases} \dfrac{\partial u_i}{\partial x_j} = u_{ij}, \\[2mm] \dfrac{\partial u_{11}}{\partial x_2} - \dfrac{\partial u_{12}}{\partial x_1} = 0, \; \dfrac{\partial u_{12}}{\partial x_2} - \dfrac{\partial u_{22}}{\partial x_1} = 0, \; \dfrac{\partial u_{13}}{\partial x_2} - \dfrac{\partial u_{23}}{\partial x_1} = 0, \end{cases}$$

sequence allows one, at least apparently to distinguish inequivalent systems from each other. The two systems, (I) and (II), of your letter, in a certain sense, do not have the same degree of generality since the solution of the first one is

$$u_1 = \frac{\partial F}{\partial x_1}, \; u_2 = \frac{\partial F}{\partial x_2}, \; u_3 = \frac{\partial F}{\partial x_3}$$

while that of the second one is

$$u_1 = \frac{\partial F}{\partial x_1} + \frac{\partial \phi}{\partial x_1}, \; u_2 = \frac{\partial F}{\partial x_2} - \frac{\partial \phi}{\partial x_2}, \; u_3 = \frac{\partial F}{\partial x_3},$$

F being an arbitrary function of x_1, x_2, x_3 and ϕ an arbitrary function of x_1, x_2. If one sticks to the previous formulae, the general solution of (I) depends on one arbitrary function of 3 variables, that of (II) on one arbitrary function of 3 variables *and* one arbitrary function of 2 variables (both statements being, moreover, in disagreement with the values of $n - r_2 + r_1$ and $n - r_1$!).

So I'm willing to agree with you that the first non-zero number of the sequence $n - r + r_2$, $n - r_2 + r_1$, $n - r_1$ is totally insufficient to give a precise idea of that fugitive thing called the degree of generality of the solution of a system. Why did I restrict myself to this single number? It's because the others are not always retained when the system is replaced by an equivalent one. I call two systems with the same number of independent variables equivalent when there exists a one-to-one correspondence between the solutions of these two systems. Take, for example, your system I

$$\frac{\partial u_3}{\partial x_2} - \frac{\partial u_2}{\partial x_3} = 0, \; \frac{\partial u_1}{\partial x_3} - \frac{\partial u_3}{\partial x_1} = 0, \; \frac{\partial u_2}{\partial x_1} - \frac{\partial u_1}{\partial x_2} = 0,$$

with 3 functions u_1, u_2, u_3 of x_1, x_2, x_3. I now consider the following system with 9 unknown functions u_i and $u_{ji} = u_{ji}$ $(i, j = 1, 2, 3)$.

$$\begin{cases} \dfrac{\partial u_i}{\partial x_j} = u_{ij}, \\[2mm] \dfrac{\partial u_{11}}{\partial x_2} - \dfrac{\partial u_{12}}{\partial x_1} = 0, \; \dfrac{\partial u_{12}}{\partial x_2} - \dfrac{\partial u_{22}}{\partial x_1} = 0, \; \dfrac{\partial u_{13}}{\partial x_2} - \dfrac{\partial u_{23}}{\partial x_1} = 0, \end{cases}$$

221

$$\left\{ \begin{array}{l} \dfrac{\partial u_{11}}{\partial x_3} - \dfrac{\partial u_{13}}{\partial x_1} = 0, \ \dfrac{\partial u_{12}}{\partial x_3} - \dfrac{\partial u_{23}}{\partial x_1} = 0, \ \dfrac{\partial u_{13}}{\partial x_3} - \dfrac{\partial u_{33}}{\partial x_1} = 0, \\[4mm] \qquad\qquad \dfrac{\partial u_{22}}{\partial x_3} - \dfrac{\partial u_{23}}{\partial x_2} = 0, \ \dfrac{\partial u_{23}}{\partial x_3} - \dfrac{\partial u_{33}}{\partial x_2} = 0. \end{array} \right.$$

Il comporte $9 + 8 = 17$ équations; il est d'autre part en involution. Il est clair que ces deux systèmes sont équivalents, les u_{ij} n'étant autres que les dérivées partielles des fonctions u_i du premier système; il y a correspondance biunivoque entre les solutions de ces deux systèmes. Pour le second on a

$$n = 9, \ r = 17, \ r_1 = 3, \ r_2 = 9;$$

on a en effet les $r_1 = 3$ équations

$$\frac{\partial u_1}{\partial x_1} = u_{11}, \frac{\partial u_2}{\partial x_1} = u_{12}, \frac{\partial u_3}{\partial x_1} = u_{13}$$

et les $r_2 - r_1 = 6$ équations

$$\frac{\partial u_1}{\partial x_2} = u_{12}, \frac{\partial u_2}{\partial x_2} = u_{22}, \frac{\partial u_3}{\partial x_2} = u_{23},$$

$$\frac{\partial u_{11}}{\partial x_2} - \frac{\partial u_{12}}{\partial x_1} = 0, \frac{\partial u_{12}}{\partial x_2} - \frac{\partial u_{22}}{\partial x_1} = 0, \frac{\partial u_{13}}{\partial x_2} - \frac{\partial u_{23}}{\partial x_1} = 0.$$

La suite $\qquad n - r + r_2, \ \ n - r_2 + r_1, \ \ \ \ n - r_1,$

est donc ici $\qquad\quad 1 \qquad\qquad 3 \qquad\qquad 6$

au lieu de $\qquad\quad\, 1 \qquad\qquad 2 \qquad\qquad 3$

suite trouvée pour le système I. Le premier nombre non nul, à savoir 1, n'a pas varié, mais les autres ont varié.

En prenant pour nouvelles fonctions inconnues les dérivées *secondes* de u_1, u_2, u_3 (en même temps que les dérivées premières), on arriverait à un nouveau système en involution pour lequel la suite serait

$$1 \qquad\qquad 4 \qquad\qquad 10$$

et ainsi de suite.

Si l'on voulait donner une suite de nombres vraiment invariante, il faudrait peut-être considérer, parmi tous les systèmes en involution équivalents entre eux, celui ou ceux qui comportent le

$$\begin{cases} \dfrac{\partial u_{11}}{\partial x_3} - \dfrac{\partial u_{13}}{\partial x_1} = 0, & \dfrac{\partial u_{12}}{\partial x_3} - \dfrac{\partial u_{23}}{\partial x_1} = 0, & \dfrac{\partial u_{13}}{\partial x_3} - \dfrac{\partial u_{33}}{\partial x_1} = 0, \\[3mm] & \dfrac{\partial u_{22}}{\partial x_3} - \dfrac{\partial u_{23}}{\partial x_2} = 0, & \dfrac{\partial u_{23}}{\partial x_3} - \dfrac{\partial u_{33}}{\partial x_2} = 0. \end{cases}$$

It is composed of $9 + 8 = 17$ equations; moreover it is in involution. It is clear that these two systems are equivalent, the u_{ij} are nothing other than the partial derivatives of the functions u_i of the first system. There is a one-to-one correspondence between the solutions of these two systems. For the second system one has

$$n = 9, \; r = 17, \; r_1 = 3, \; r_2 = 9.$$

Thus one has the $r_1 = 3$ equations

$$\frac{\partial u_1}{\partial x_1} = u_{11}, \quad \frac{\partial u_2}{\partial x_1} = u_{12}, \quad \frac{\partial u_3}{\partial x_1} = u_{13}$$

and the $r_2 - r_1 = 6$ equations

$$\frac{\partial u_1}{\partial x_2} = u_{12}, \quad \frac{\partial u_2}{\partial x_2} = u_{22}, \quad \frac{\partial u_3}{\partial x_2} = u_{23},$$

$$\frac{\partial u_{11}}{\partial x_2} - \frac{\partial u_{12}}{\partial x_1} = 0, \quad \frac{\partial u_{12}}{\partial x_2} - \frac{\partial u_{22}}{\partial x_1} = 0, \quad \frac{\partial u_{13}}{\partial x_2} - \frac{\partial u_{23}}{\partial x_1} = 0.$$

So, the sequence $\quad n - r + r_2, \quad n - r_2 + r_1, \quad n - r_1,$

is here $\qquad\qquad\qquad 1 \qquad\qquad\quad 3 \qquad\qquad\quad 6,$

instead of $\qquad\qquad\quad 1 \qquad\qquad\quad 2 \qquad\qquad\quad 3,$

the sequence found for the system I. The first non-zero number, namely 1, has not changed, but the others have.

  If one took, as new unknown functions, the *second* derivatives of u_1, u_2, u_3 (together with the first derivatives), one would have a new system in involution, for which the sequence would be

$$1 \qquad\qquad 4 \qquad\qquad 10$$

and so on.

  If one wanted to give a really invariant sequence of numbers, perhaps one might have to consider, among all the systems in involution equivalent to each other, the one or several that contain the smallest number of unknown functions; and still, one would not be

moindre nombre de fonctions inconnues, et encore n'est-il pas sûr que deux systèmes équivalents au nombre minimum de fonctions inconnues donneraient naissance à la même suite; du moins faudrait-il le démontrer. Mais tout cela conduirait à des problèmes dont la difficulté me semble hors de proportion avec l'intérêt que présente la question.

Le nombre que j'ai introduit a un sens vraiment invariant (toutes les fois qu'on le calcule en partant d'un choix *non singulier* des variables), et dans beaucoup de questions la considération de ce nombre suffit pour prouver la non-équivalence de deux systèmes donnés, ou encore par exemple pour résoudre des questions analogues à l'impossibilité de mettre une forme différentielle quadratique arbitraire à 4 variables sous la forme d'une somme de 4 carrés $H_1 dx_1^2 + H_2 dx_2^2 + H_3 dx_3^2 + H_4 dx_4^2$, etc.

Après cette trop longue lettre, j'ose espérer que si nous ne sommes pas encore tout à fait d'accord, nous aurons du moins rapproché les distances! J'ai été vraiment touché de ce que sur une chose qui, comme vous le dites, est assez insignifiante, vous n'ayez pas eu de repos avant de vous être mis d'accord avec moi.

Je vous prie de croire, cher et illustre Maître, à toute ma cordiale affection.

E. Cartan

sure that two equivalent systems with the minimum number of unknown functions would give the same sequence; or, at least, one would have to prove it. But all that would lead to difficult problems out of proportion to the interest of the question.

The number I introduced really has an invariant meaning (each time one computes it starting from a *non-singular* choice of the variables), and for many questions this number is sufficient to prove that two systems are not equivalent, or to solve, for example, questions analogous to the impossibility of writing an arbitrary quadratic differential form in 4 variables as the sum of 4 squares

$$H_1 dx_1^2 + H_2 dx_2^2 + H_3 dx_3^2 + H_4 dx_4^2, \text{ etc.}$$

After this too long letter, I venture to hope that if we do not yet completely agree, we will at least have shortened the distance between us. I was really touched that on a subject which, as you say, is rather unimportant, you will have had no rest until you came to an agreement with me.

E. Cartan

XXXVIII

16-V-32

Verehrter und lieber Herr Cartan!

Nach einem so sorgfältigen und ausführlichen Briefe von Ihnen greife ich zur Feder mit einem schlechten Gewissen, weil es ein Skandal ist, wenn ich Ihre grosse Güte weiter missbrauche. Aber ich tröste mich mit der Illusion: vielleicht hat doch auch *er* daneben ein bischen Freude an dieser kleinen Auseinandersetzung. Denken Sie also, wir seien wieder beide jung und ich sei ein zwar eifriger aber leider Schüler von Ihnen.

Zuerst hat mich Ihr Brief überzeugt und Ihr Kompromiss erschien auch mir vorläufig das Beste. Dann aber kamen Zweifel, die ich nicht mehr los werden konnte.

Die Gleichung

$$\frac{\partial^2 z}{\partial y^2} - \frac{\partial z}{\partial x} = 0$$

ist bezüglich der Fortsetzung (Entwicklung) nach x von der ersten Ordnung. Davon kommt es, dass man bei dieser Entwicklung mit *einer* willkürlichen Funktion (von y) auskommt. Es scheint mir nun ausgemacht, dass man bei der andern Entwicklungsart nur *scheinbar* zwei willkürliche Funktionen in die allgemeine Lösung hineinbekommt, dass sich diese aber eben in *eine* zusammenfassen lassen. Hieran ist eigentlich gar nicht zu zweifeln. Es handelt sich eben um eine Degeneration der partiellen Gleichungen zweiter Ordnung, wenn $\frac{\partial}{\partial x}$ in den zweimal abgeleiteten Gliedern nicht auftritt, wohl aber in den einmal agbeleiteten. Dies ist sehr überraschend für mich, scheint mir aber ohne Einfluss zu sein auf die Beantwortung der von uns ins Auge gefassten Frage.

Letzeres scheint mir auch zuzutreffen für diejenige Bemerkung, welche an die Gleichung

$$F(x) = f(x^2) + x\phi(x^2)$$

angeknüpft ist. Die beiden willkürlichen Funktionen f und ϕ bestimmen eben zusammen nur *eine* willkürliche Funktion von x.

226

16 MAY 1932

A XXXVIII

16.V.32

My very dear M. Cartan!

After such a careful and detailed letter from you I take up my pen with a heavy conscience; for it is a shame for me to further misuse your great gifts. But I console myself with the illusion: perhaps *he* too obtains a bit of joy from this little discussion. So imagine we are both young again and I am a keen but troublesome student of yours.

At first your letter convinced me and your compromise also seemed, provisionally, the best one. But then came doubts which I could not shake off.

The equation

$$\frac{\partial^2 z}{\partial y^2} - \frac{\partial z}{\partial x} = 0$$

is of first order with respect to the continuation (expansion) in x. From this it follows that this expansion can be represented by a *single* arbitrary function (of y). Now it seems settled to me that another kind of expansion only *apparently* brings two arbitrary functions into the general solution, but that really these can be put together into *one*. Here there really is no doubt. It is simply a question of the degeneration of a partial differential equation of the second order when $\frac{\partial}{\partial x}$ does not appear in the second derivative terms but only in the first derivative ones. This is very surprising to me thought it seems to be without influence on the answer to the question before us.

This last also appears to me to be true in the case of the equation

$$F(x) = f(x^2) + x\phi(x^2).$$

Both arbitrary functions, f and ϕ, together really appear to fix just *one* arbitrary function of x.

227

Nun kommt aber die Hauptsache. Sie geben für das Gleichungssystem

$$\frac{\partial u_3}{\partial x_1} - \frac{\partial u_1}{\partial x_3} = 0,$$

$$\frac{\partial u_3}{\partial x_2} - \frac{\partial u_2}{\partial x_3} = 0,$$

$$\vdots \qquad \vdots$$

die Lösung

$$u_1 = \frac{\partial F(x_1,x_2,x_3)}{\partial x_1} + \frac{\partial \phi(x_1,x_2)}{\partial x_1},$$

$$u_2 = \frac{\partial F(x_1,x_2,x_3)}{\partial x_2} + \frac{\partial \phi(x_1,x_2)}{\partial x_2},$$

$$u_3 = \frac{\partial F(x_1,x_2,x_3)}{\partial x_3}.$$

Diese Lösung ist aber nicht die *allgemeine* Lösung. Sie ist ja sogar enthalten in der allgemeinen Lösung des Systems $\frac{\partial u_\mu}{\partial x_\nu} - \frac{\partial u_\nu}{\partial x_\mu} = 0$ (μ und ν von $1-3$). Setzt man

$$F + \phi = F^*,$$

so ist Ihre Lösung

$$u_\nu = \frac{\partial F^*}{\partial x_\nu}.$$

Es durf also hieraus kein Schluss über die Beschaffenheit der allgemeinen Lösung des ins Auge gefassten Gleichungssystems gezogen werden.

Ihr letztes Beispiel habe ich noch nicht genau durchgeprüft.

Es wird Ihnen nun komisch vorkommen, dass ich immer noch auf dieser Bagatelle herumreite. Dahinter steckt aber doch die allgemeine Frage, ob man derartige Mannigfaltigkeiten bezüglich ihrer „ Mächtigkeit ", irgendwie klar und eindeutig charakterisieren kann oder nicht, und diese Frage scheint mir eben doch interessant zu sein.

Mit herzlicher Bitte um Nachricht und Geduld bin ich freundlich grüssend

Ihr

A. Einstein

228

But now we come to the main point. For the system of equations

$$\frac{\partial u_3}{\partial x_1} - \frac{\partial u_1}{\partial x_3} = 0,$$

$$\frac{\partial u_3}{\partial x_2} - \frac{\partial u_2}{\partial x_3} = 0,$$

$$\vdots \qquad \vdots$$

you give the solution:

$$u_1 = \frac{\partial F(x_1,x_2,x_3)}{\partial x_1} + \frac{\partial \phi(x_1,x_2)}{\partial x_1},$$

$$u_2 = \frac{\partial F(x_1,x_2,x_3)}{\partial x_2} + \frac{\partial \phi(x_1,x_2)}{\partial x_2},$$

$$u_3 = \frac{\partial F(x_1,x_2,x_3)}{\partial x_3}.$$

However, this solution is not the *general* solution. What is more, it is contained in the general solution of the system $\dfrac{\partial u_\mu}{\partial x_\nu} - \dfrac{\partial u_\nu}{\partial x_\mu} = 0$ (μ and ν run from $1 - 3$). Putting

$$F + \phi = F^*,$$

your solution becomes

$$u_\nu = \frac{\partial F^*}{\partial x_\nu}.$$

Therefore, no conclusion should be drawn from this as to the nature of the general solution of the system of equations in question.

I have not yet thoroughly examined your last example.

Now it will appear strange to you that I constantly harp on this bagatelle. But behind this there is the general question of whether one can or not characterize manifolds of such a kind by their "strength" in some clear and unique way; and this question, it seems to me, is very interesting.

With a heartfelt request for patience and indulgence and with friendly regards.

Yours,

A. Einstein

XXXIX

Le Chesnay, le 24 mai 1932

Cher et illustre Maître,

Votre lettre me remplit à la fois de joie et de confusion. Sûrement j'éprouve du plaisir à notre petite correspondance; s'il ne tenait qu'à moi je redeviendrais volontiers jeune, sinon pour vous donner des leçons, du moins pour pouvoir, mieux que je ne le puis maintenant, suivre tout ce qui se fait de merveilleux en physique.

Mais revenons à nos équations aux dérivées partielles. J'ai dû commettre un lapsus dans ma lettre au sujet du système $\frac{\partial u_3}{\partial x_1} - \frac{\partial u_1}{\partial x_3} = 0, \frac{\partial u_3}{\partial x_2} - \frac{\partial u_2}{\partial x_3} = 0$ dont la solution générale est

$$u_1 = \frac{\partial F(x_1, x_2, x_3)}{\partial x_1} + \frac{\partial \phi(x_1, x_2)}{\partial x_1},$$

$$u_2 = \frac{\partial F(x_1, x_2, x_3)}{\partial x_2} - \frac{\partial \phi(x_1, x_2)}{\partial x_2},$$

$$u_3 = \frac{\partial F(x_1, x_2, x_3)}{\partial x_3},$$

les fonctions F et ϕ n'étant pas du reste complètement définies pour une solution donnée (on peut ajouter à F la somme $X_2 + X_1$ et ajouter à ϕ la différence $X_2 - X_1$, où X_1 ne dépend que de x_1 et X_2 que de x_2).

Quant au problème général lui-même je ne vois guère, comme je vous le disais, d'autre manière d'en donner une solution raisonnable qu'en recourant au système équivalent qui comporte le nombre minimum de fonctions inconnues, et encore je ne suis pas sûr ainsi d'arriver à un résultat satisfaisant.

Bien affectueusement

E. Cartan

A XXXIX

Le Chesnay, 24 May 1932

Cher et illustre Maître,

Your letter has filled me with both joy and confusion. Of course I take pleasure in our little exchange and, if it were up to me, I would willingly become young again, if not to give you lessons, at least to follow, better than I now can, all the marvelous things being done in physics.

But let us return to our partial differential equations. I must have made a slip in my letter on the subject of the system

$$\frac{\partial u_3}{\partial x_1} - \frac{\partial u_1}{\partial x_3} = 0, \frac{\partial u_3}{\partial x_2} - \frac{\partial u_2}{\partial x_3} = 0$$

whose general solution is

$$u_1 = \frac{\partial F(x_1, x_2, x_3)}{\partial x_1} + \frac{\partial \phi(x_1, x_2)}{\partial x_1},$$

$$u_2 = \frac{\partial F(x_1, x_2, x_3)}{\partial x_2} - \frac{\partial \phi(x_1, x_2)}{\partial x_2},$$

$$u_3 = \frac{\partial F(x_1, x_2, x_3)}{\partial x_3},$$

Moreover, the functions F and ϕ are not completely defined for a given solution (one can add the sum $X_2 + X_1$ to F and add the difference $X_2 - X_1$ to ϕ, where X_1 depends only on x_1 and X_2 only on x_2).

As for the general problem itself, as I have said, I see no other way of a reasonable solution for it than by having recourse to the equivalent system having the minimum number of unknown functions. And even so, I am not sure of arriving at a satisfactory result.

With affection,

E. Cartan

Bibliographie

[1] CARTAN, E. Sur certaines expressions différentielles et le problème de Pfaff. *Ann. Ec. Normale*, t. 16, pp. 239-332, 1899. *O.C.*, II, 1, pp. 303-396. *O.C.* désigne les *Œuvres complètes* d'Élie Cartan, Gauthier-Villars, Paris, 1952-1955.

[2] CARTAN, E. Sur l'intégration des systèmes d'équations aux différentielles totales. *Ann. Ec. Normale*, t. 18, pp. 241-311, 1901. *O.C.*, II, 1, pp. 411-481.

[3] CARTAN, E. Sur une généralisation de la notion de courbure de Riemann et les espaces à torsion. *C.R.Acad. Sc. Paris*, t. 174, pp. 593-595, 1922. *O.C.*, III, 1, pp. 616-618.

4] CARTAN, E. Sur les variétés à connexion affine et la théorie de la relativité généralisée. *Ann. Ec. Norm.*, t. 40, pp. 325-412, 1923. *O.C.*, III, 1, pp. 659-746; t. 41, pp. 1-25, 1924. *O.C.*, III, 1, pp. 799-823; t. 42, pp. 17-88, 1925, *O.C.* III, 2, pp. 921-992.

[5] CARTAN, E. Les récentes généralisations de la notion d'espace. *Bull. Sc. math.* t. 48, pp. 294-320, 1924. *O.C.*, III, 1, pp. 863-889.

[6] CARTAN, E. La théorie des groupes et les recherches récentes de géométrie différentielle. (Conf. Congrès Int. de Math. Toronto, août 1924.) *L'Enseignement math.*, t. 24, 1-18, 1925. *O.C.* III, 1, pp. 891-904.

[7] CARTAN, E. La géométrie des groupes de transformations. *J. de Math. pures et appliquées*, t. 6, pp. 1-119, 1927, *O.C.*, I, 2, pp. 673-791.

[8] CARTAN, E. La théorie des groupes et la géométrie. *L'Enseignement math.*, t. 26, pp. 200-225, 1927, *O.C.*, I, 2, pp. 841-866.

[9] CARTAN, E. Notice historique sur la notion de parallélisme absolu. *Math. Annalen*, 102, pp. 698-706, 1930. *O.C.*, III, 2, pp. 1121-1129.

[10] CARTAN, E. La théorie des groupes finis et continus et l'*Analysis situs*. *Mémorial Sc. Math.* XLII. Gauthier-Villars, Paris, 1930. *O.C.*, I, 2, pp. 1165-1225.

[11] CARTAN, E. Géométrie euclidienne et géométrie riemannienne. *Scientia*, 49, pp. 393-402, 1931.

[12] CARTAN, E. Le parallélisme absolu et la théorie unitaire du champ. *Revue de Métaphysique et de Morale*, 38, pp. 13-28, 1931. *Act. sc. et ind.* XLIV, Hermann, Paris, 1932. *O.C.*, III, 2, pp. 1167-1185. Voir [14]

[13] CARTAN, E. Sur la théorie des systèmes en involution et ses applications à la relativité. *Bull. Soc. Math. France*, t. 59, pp. 88-118, 1931. *O.C.*, II, 2, pp. 1199-1229.

[14] CARTAN, E. Notice sur les Travaux Scientifiques, datée de 1931, publiée aux *Selecta*, Jubilé Scientifique de M. Elie Cartan. Gauthier-Villars, Paris 1939, pp. 15-112, *O.C.*, I, 1, pp. 1-98. Gauthier-Villars a, en 1974, dans la

232

collection « Discours de la Méthode », reproduit cette notice et joint à la suite le texte de [12].

[15] CARTAN, E. La théorie unitaire d'Einstein-Mayer. (1934?). *O.C.*, III, 2, pp. 1863-1875.

[16] EINSTEIN, A. Riemann-Geometrie mit Aufrechterhaltung des Begriffes des Fernparallelismus. *Sitzungsberichte. Preuss. Akad. Wiss.* (abrévié Stz.) pp. 217-221, séance du 7 juin 1928.

[17] EINSTEIN, A. Neue Möglichkeit für eine einheitliche Feldtheorie von Gravitation und Elektrizität. *Stz.*, pp. 224-227, 14 juin 1928.

[18] EINSTEIN, A. Zur einheitlichen Feldtheorie. *Stz.* pp. 2-8, 10 janvier 1929. Il en existe une traduction française sous le titre : « Sur la théorie synthétique des champs » dans *Revue générale d'électricité*, t. XXV, n° 17, 27 avril 1929, pp. 644-648, précédé d'une courte introduction de Th. De Donder.

[19] EINSTEIN, A. Einheitliche Feldtheorie und Hamiltonsches Prinzip. *Stz.*, pp. 156-159, 21 mars 1929.

[20] EINSTEIN, A. Auf die Riemann-Metrik und den Fern-Parallelismus gegründete einheitliche Feldtheorie. *Math. Annalen*, Bd 102, pp. 685-697, 1930.

[21] EINSTEIN, A. Théorie unitaire du champ physique. *Annales de l'Institut H. Poincaré*, I, pp. 1-24, 1930.

[22] EINSTEIN, A. Die Kompatibilität der Feldgleichungen in der einheitlichen Feldtheorie. *Stz.*, pp. 18-23, 12 décembre 1929.

[23] EINSTEIN, A. Zwei strenge statische Lösungen der Feldgleichungen der einheitliche Feldtheorie. (en coll. avec W. Mayer). *Stz.*, pp. 110-120, 20 février 1930.

[24] EINSTEIN, A. et MAYER, W. Systematische Untersuchung über kompatible Feldgleichungen, welche in einem Riemannschen Raume mit Fernparallelismus gesetzt werden können. *Stz.*, pp. 257-265, 23 avril 1931.

[25] EINSTEIN, A. et MAYER, W. Einheitliche Theorie von Gravitation und Elektrizität. *Stz.*, pp. 541-557, 22 octobre 1931.

[26] EINSTEIN, A. et MAYER, W. Einheitliche Theorie von Gravitation und Elektrizität (Zweite Abhandlung). *Stz.*, pp. 130-137, 14 avril 1932.

[27] EISENHART, L. P. Non-Riemannian Geometry. *Amer. Math. Soc. Colloq. Publications* VIII, New York, 1927.

[28] GROSSMANN, M. Fernparallelismus? *Vierteljahrsschrift der Naturforschenden Gesellschaft in Zürich.* 76, pp. 42-60, 1931 (Mss. du 19 janvier 1931).

[29] WEITZENBÖCK, R. Differentialinvarianten in der Einsteinschen Theorie des Fernparallelismus. *Stz.*, pp. 466-474, 18 octobre 1928.

CONTENTS

Plates:

Letter by E. CARTAN, 3 December 1929, front of p. 22.
Post-Card by A. EINSTEIN, 11 January 1930, front of p. 122.

Imprimerie J. Duculot - Gembloux - Belgique